Canto is a paperback imprint
which offers a broad range of titles,
both classic and more recent,
representing some of the best
and most enjoyable of Cambridge
publishing.

Seven clues to the origin of life

a scientific detective story

A. G. CAIRNS-SMITH

CAMBRIDGE
UNIVERSITY PRESS

PUBLISHED BY THE PRESS SYNDICATE OF THE UNIVERSITY OF CAMBRIDGE
The Pitt Building, Trumpington Street, Cambridge, United Kingdom

CAMBRIDGE UNIVERSITY PRESS
The Edinburgh Building, Cambridge CB2 2RU, UK   http://www.cup.cam.ac.uk
40 West 20th Street, New York, NY 10011–4211, USA   http://www.cup.org
10 Stamford Road, Oakleigh, Melbourne 3166, Australia
Ruiz de Alarcón 13, 28014 Madrid, Spain

© Cambridge University Press 1985

First published 1985
First paperback edition 1986
Reprinted 1987, 1988
Canto edition 1990
Reprinted 1991, 1995, 1998, 2000

Printed in the United Kingdom at the University Press, Cambridge

Library of Congress catalogue card number: 84-23044

British Library Cataloguing in Publication Data
Cairns-Smith, A. G.
Seven clues to the origin of life: a scientific detective story.
1. Life – Origin
I. Title
577   QH325

ISBN 0 521 39828 2 paperback

Cover illustration: Diatoms (Photo: Jan Hinsch/Science Photo Library)

To Sarah, Adam and Emma

CONTENTS

PREFACE

Faced with a really difficult-looking problem should one follow the advice of Descartes or of Holmes? Should one proceed step by step from what is easily understood, as Descartes advised, 'starting with what was simplest and easiest to know, and rising little by little to the knowledge of the most complex'? It sounds like good advice and on the whole modern science takes it. But the methodical step-by-step strategy does not always work. First steps can be particularly tricky and you have to know in which direction to go. There are times when you need the advice not of Descartes, but of Sherlock Holmes.

You see Holmes, far from going for the easy bits first, would positively seek out those features in a case that were seemingly incomprehensible – 'singular' features he would call them. They can point the way. They can tell you what sort of a problem it is that you are dealing with. If you can see how the murder could have been done *at all* with the door and windows securely fastened..., or if you can understand why on earth the thief should have rung the bell that gave away his presence in the room..., why then, you may even have cracked the whole thing.

I think that the origin of life is a Holmesian problem – that if we can understand how life could have started *at all*, then we should be able to work out, roughly at least, how it *did* start.

Much of this book is devoted to seeking out, and making as stark as possible, the difficulties in the case of the origin of life on the Earth. Not so that we can throw up our hands and say 'Look how impossible it all is!' Not at all. Rightly or wrongly we will be assuming that life really did arise on this Earth 'from natural causes'. We look for difficulties to see as clearly as possible what the real problem is, and to fashion a key to unlock it.

Seven Clues... started from an idea to write a layman's version of my book *Genetic Takeover* – something much shorter, with only a few technical terms and diagrams, and no references. Since in any case the problem of the origin of life is one that calls for detective work, I thought it would be amusing to write the new book somewhat in the style of a detective story. You can read it like this if you want to, trying to foresee the curious conclusion that will start to emerge around chapter 10.

There will be plenty of other implicit questions for you to think about: What are the real difficulties? Which are the main suspects? Where are the red herrings? What are the best clues? (Or anyway what does the author think they are?) My choice of the seven best clues, dropped at various stages in the book, will be listed and located explicitly in the final chapter.

I decided to have no references because (i) laymen are not particularly interested in who did what when, (ii) the *cognoscenti* know pretty well anyway, and then (iii) because I had already written a book that was full of references. So you will find no more than a little light name-dropping here and there. For laymen I need only stress that not all the ideas are original by any means: even those that have some originality are derived from earlier ideas, and in any case the whole background of knowledge that allows one even to start speculating about our origins depends on innumerable experiments and observations by others. Then again ideas have been formed, sharpened – or thrown out – as a result of very many discussions with friends and colleagues over many years.

I would like to thank in particular those who actively helped in the preparation of this book, by reading the manuscript and discussing it with me: Paul Braterman, Colin Brown, Roger Buick, Jack Cohen, John Freer, Sally Gibson, Hyman Hartman and Kelvin Tyler, as well as my wife Dorothy Anne and my son Adam. I am grateful also to Janet McIntyre and to my daughter Sarah who got the words smoothly onto the typescript by various electronic means.

Glasgow, Spring 1984   Graham Cairns-Smith

SOURCES OF QUOTATIONS

Quotations at the start and finish of each chapter are from the Sherlock Holmes stories of Sir Arthur Conan Doyle.

(A, *Adventures*; C, *The Case Book*; L, *His Last Bow*; M, *Memoirs*; R, *The Return*.)

1

Inquest

'Do you see any prospect of solving this mystery, Mr. Holmes?' she asked
with a touch of asperity in her voice.
'Oh, the mystery!' he answered, coming back with a start to the realities
of life. 'Well, it would be absurd to deny that the case is a very abstruse
and complicated one, but I can promise you that I will look into the
matter and let you know any points which may strike me.'
'Do you see any clue?'
'You have furnished me with seven, but of course I must test them
before I can pronounce upon their value.'
'You suspect someone?'
'I suspect myself.'
'What!'
'Of coming to conclusions too rapidly.'

Whatever one might read in the newspapers there is not much real
distress among biologists about the fundamental idea of biology –
which is **evolution**. There may be arguments still about how, and how
fast, evolutionary changes took place. But that evolution did take
place is hardly an issue any more. The idea that the multitudinous
forms of life on the Earth have evolved from common ancestors owes
its security not to some single demonstration but to a more day-to-day
experience of biologists – that this idea fits with innumerable detailed
and general observations. In providing a global view of the relatedness
of life it is an idea that makes biology into a coherent subject. Biology
has become, quite simply, the study of the causes and effects of
evolution, and the question of the origin of life is, first, the question
of the origin of evolution.

But to frame the matter in this rather stolid way is not to deny that
the case of the origin of life on the Earth is an abstruse and
complicated one. There are clues certainly, many more than Holmes'
seven, but their significance is indirect and the most obvious are not
always the most important. We will be thought-testing many of these
clues, detecting a number of red herrings, and avoiding coming to
conclusions too rapidly. What will emerge will be seven clues that
seem to be of particular use, and an overall view of the origin of life
that these clues suggest.

Before we can really get started there are some words to disentangle.
First, there is the word **life**.

My dictionary tells me that life is the period between birth and death – but that is not what I want to talk about. This book is about life as a phenomenon – as in 'life on the Earth'. Life is a property that is shared by man, moulds and marigolds. It is a rather fuzzy property, unfortunately, notoriously difficult to pin down, even if obvious enough in most cases.

I prefer uses of the word 'life' that admit the fuzziness, that do not pin down but somehow encompass the general idea. Coleridge's definition is like this: 'I define life as...a whole that is pre-supposed by all its parts'. After all what impresses us about a living thing is its in-built ingenuity, its appearance of having been designed, thought out – of having been put together with a purpose. Life can be thought of as a kind of naturally occurring machinery. We can see what the purpose of a living thing is: it is to survive, to compete, to reproduce its kind against the odds.

Coleridge, it must be admitted, was looking for something more poetic than machinery – for some deeply mysterious unifying power, a principle of life, an essential magic about living things that divides them from everything else. That is what we call vitalism. It is officially outmoded; but there are still respected scientists, physicists mainly, who seem to want of 'life' more than machinery: who would seek some profound division. There is a temptation, in any case, to suppose that if the origin of life was not actually supernatural it was at least some very extraordinary event, an event of low probability, a statistical leap across a great divide. That way a trace of magic can be held on to.

I am much more inclined to an opposite prejudice – a majority prejudice now. This is to suppose that the exorcism that Darwin initiated will continue right back to the origin of life.

Darwin persuades us that the seemingly purposeful construction of living things can very often, and perhaps always, be attributed to the operation of **natural selection**. *If* you have things that are reproducing their kind; *if* there are sometimes random variations, nevertheless, in the offspring; *if* such variations can be inherited; *if* some such variations can sometimes confer an advantage on their owners; *if* there is competition between the reproducing entities – *if* there is an overproduction so that not all will be able to survive to produce offspring themselves – then these entities will get better at reproducing their kind. Nature acts as a selective breeder in these circumstances: the stock cannot help but improve.

You can see that natural selection is a good bit more than just 'survival of the the fittest': there are all those 'ifs'. And natural selection is only one component of the mechanism of evolution. Any theory that is to explain the variety and complexity of living things must also take into account the varied and varying challenges set up by a varied and varying environment. Nature, as breeder and show judge, is continually changing her mind about which types should be awarded first prize: changing selection pressures have been a key part of her inventiveness.

But nevertheless natural selection has been *the* key component, the *sine qua non*. Without it, living things could not even stay adapted to a given set of circumstances, never mind become adapted to new ones. Without natural selection the whole adventure would never have got off the ground. That kind of in-built ingenuity that we call 'life' is easily placed in the context of evolution: *life is a product of evolution*.

Life, then, is not some absolute quality that would have suddenly appeared, it would have emerged gradually during early evolution. Between the first evolving entities, that first conformed to those 'ifs', and later forms that did so more cleverly, there would have been no hard line. But who cares? When you can see what is going on in such a progression you lose interest in erecting picket fences. As I said, 'life' is a fuzzy idea – and it is best left like that.

But wait a minute, you may say, do you not need living things to generate 'life' in this way?

No. What is needed for evolution is natural selection, and what is needed for natural selection are things that conform to those 'ifs'. There is nothing in the rules to say that such things had to be 'living'. We have an odd view now because all of those things that we are aware of that *can* evolve *have* evolved.

So the start of our problem seems clear enough: if we are to come to understand the origin of life – which would have been a gradual emergence – we must first understand the origin of evolution. We must first find things that can evolve but have not yet done so. If indeed there were such things that were in at the start of evolution there should still be things like them to be found or made.

Organism is another word that can be placed in the context of evolution: organisms are participants in evolution. More specifically, for our purposes, *organisms are prerequisites for evolution*. Those first evolvable things that we have just been talking about would not yet have been 'living' – but they would still have been organisms.

It might seem that by far the most difficult parts of the problem of the origin of life had already been solved in principle once there was a general understanding of the way evolution works and of the way in which organisms reproduce and pass on characteristics to offspring. Already towards the middle of this century there were biochemists and others who felt that there might be only a little gap between some bacterium of the kind that could be seen as a somewhat unstructured blob under a microscope, and some still smaller congregation of molecules that might have been a starter organism for evolution.

The early fifties were a high point of optimism. The molecular basis of heredity had suddenly been uncovered, and prevailing ideas about conditions on the primitive Earth suggested that the component parts for such molecular machinery would have been just lying about. It was thought that the atmosphere on the early Earth would have been like Jupiter's atmosphere now; that it would have been dominated by gases such as hydrogen, methane and ammonia. It had been shown that with such gases some of those amino acids that are now found in all organisms can be made fairly easily. Given such pieces there was always chance to do the rest, and who knows, there might be special effects to be discovered that would reduce the amount of luck that would be needed.

The optimism persists in many elementary textbooks. There is even, sometimes, a certain boredom with the question; as if it was now merely difficult because of an obscurity of view, a difficulty of knowing now the details of distant historical events.

What a pity if the problem had really become like that! Fortunately it hasn't. It remains a singular case (Sherlock Holmes' favourite kind): far from there being a million ways in detail in which evolution could have got under way, there seems now to have been no obvious way at all. The singular feature is in the gap between the simplest conceivable version of organisms as we know them, and components that the Earth might reasonably have been able to generate. This gap can be seen more clearly now. It is enormous.

Three prime facts

By far the most remarkable fact of this case is already known to readers:

Fact one: There is life on the Earth

That there is a profusion of forms of life will not have gone unnoticed either; but it is much less apparent that this profusion is, biochemically, somewhat superficial. If you could use a big enough magnifying glass you would find that there is really only one kind of life on the Earth: the most central machinery in all organisms is built out of the same set of micro-components, the same set of small molecules. We have then:

Fact two: All known living things are at root the same

But the fact that causes all the trouble is this one:

Fact three: All known living things are very complicated

While it may be so that several of the common micro-components are fairly simple molecules in themselves, they collaborate in a way that is both highly organised and complex. This in itself might be dismissed with a waving hand as a product of evolution (' – of course things would have been much simpler to start with –'). But the real trouble arises because too much of the complexity seems to be necessary to the whole way in which organisms work. Our kind of life is 'high-tech'. Even some of those essential micro-components are not at all easily made. We will be returning to the perplexities of fact three.

Questions of time and circumstance

The Earth is 4.5 billion (thousand million) years old. This is quite a reliable figure. The ages of many ancient rocks are reliable too, and they go back a long way: there are rocks in Greenland that are 3.8 billion years old. The first signs of life in ancient rocks are not so easily dated (or identified), but there is good evidence now for microbes of some sort having been on the Earth at least by 2.8 billion years ago. This is a cautious estimate; most experts would say that there is now pretty good evidence for life on the Earth 3.5 billion years ago. A few would put the date as far back as 3.8.

The most direct evidence is twofold. First there are rather curious large-scale structures in many ancient rocks, including 3.5-billion-year-old Australian rocks, that resemble structures (stromatolites) that today are produced by large colonies of microbes. And then second, there are objects found in ancient rocks that appear to be fossils of microbes themselves.

Moving to the other end of the time range for the origin of life on the Earth, the earliest possible date is fixed by the age of the Earth itself, but there is evidence from the Moon and other planets that the Earth was being bombarded by very large meteorites up to about 4.0 billion years ago. So the long-stop is perhaps nearer 4.0 than 4.5

billion years ago. On the other hand the (cautious) 2.8 date for the near end of the range is likely to move further away with new evidence. So, probably, the gap will narrow. But in the meantime we can be content with a time range of 4.5 (or less) to 2.8 (or more) billion years ago for the origin of life on the Earth.

As for the conditions on the Earth when life originated, the best evidence that we have is from those Greenland rocks, 3.8 billion years old. That date is well within our range for the origin of life. The rocks themselves speak of an Earth that was not so vastly different from today. These rocks used to be sediments: they were laid down under large bodies of water. And there was presumably land too, to provide the materials to be sedimented. The Greenland rocks contain carbonates – so there was presumably carbon dioxide in the atmosphere – and also iron-containing sediments of a kind that, most probably, could only have been formed if there had been little or no free oxygen in the atmosphere. And it is generally supposed that there was also nitrogen in the early atmosphere to provide the main constituent, as now.

Of that early, heavy, Jupiter-like atmosphere, full of methane, ammonia and such things, there is no evidence from ancient rocks; and there is little enthusiasm for the idea now, either among geologists, geochemists or planetary astronomers.

Finally, let us look briefly at four extreme attitudes of the sort that tend to obliterate further enquiry.

Was the law broken?

It is a sterile stratagem to insert miracles to bridge the unknown. Soluble problems often seem to be baffling to begin with. Who would have thought a thousand years ago that the size of an atom or the age of the Earth would ever be discovered? Poor Dr. Watson was always being baffled by Sherlock Holmes' cases – as we all are by a good conjuring show. It is silly to say that because we cannot see a natural explanation for a phenomenon then we must look for a supernatural explanation. (It is usually silly anyway.) With so many past scientific puzzles now cleared up there have to be very clear reasons not to presume natural causes. Let us not say that the law was broken.

Was there a freak event?

We touched on this already. The argument goes along these lines. If the Universe is infinite in time or space then any circumstance with a finite probability will happen (an infinite number of times to boot).

Hence the first organism could have appeared on the Earth by chance. That no life has been found elsewhere in the Universe can be taken as evidence in support of this contention. This is the scientifically respectable version of the special miracle. It is just as sterile. If life *could* have started without a freak event then that is the way that it *would* have started.

Was there a frame-up?

Then there is an idea, developed particularly by Brandon Carter, called 'the anthropic principle'. We must be in a Universe that made our existence possible, so the Universe that we are in may very well be a freak Universe self-selected from myriads of others – or some might say specially contrived by God. According to this idea the laws and constants of Nature, and/or the initial conditions for the Universe, were in effect tuned to allow the production of conscious beings. Maybe so. But as with the freak event idea, there is no real let-out. However contrived our Universe may be we should be looking for the least improbable way in which life could have started within that Universe.

Was it an outside job?

Perhaps life did not start on the Earth, but seeds of some sort arrived from elsewhere.

In the early part of this century the Swedish chemist Svante Arrhenius speculated that the pressure of light waves could have pushed spores from one planetary system to another. Alternatively meteorites have sometimes been suggested as carriers, buried spores being thus protected, on their journeys, from destructive radiation.

There are practical questions of whether in fact spores would survive such journeys or be likely to find another locale suited to them. The idea suffers too from the objection that the problem is simply being displaced. At least we know that life can thrive on the Earth – on the face of it the Earth then seems as good a place as any to try to imagine the origin of life.

More recently Fred Hoyle and Chandra Wickramasinghe have speculated that in the vast reaches of interstellar space conditions may be particularly suited to the origin of life. Space then might be full of spores that had never been on a planet. Perhaps, though, such space-borne organisms could nevertheless infect a planet such as the Earth. This time the problem is being displaced more radically, by introducing the idea that perhaps space would be a better place for the critical initial phases of an evolutionary process. But that point is not at all clear; and it is not clear either that organisms that had evolved in space would survive under the very different conditions of a planetary surface. There are similar attractions and snags about the notion that comets or meteorites might have been starting places for evolution.

Francis Crick and Leslie Orgel have thought about a third idea: directed panspermia. Perhaps we are the outcome of a research project of some other intelligent beings; perhaps the Earth was deliberately seeded at some time in the remote past. Again there is a displacement here: How did the intelligent beings evolve?

Verdict

Now I cannot deny all these possibilities: life on the Earth may be a miracle, or a freak, or an alien infection. And I agree that the confidence was misplaced that supposed in the fifties that the answer to the origin of life would appear in some footnote to the answer to the question of how organisms work. Something much more will be needed. Something odd.

It is a difficult judgement at such times to say which speculations are reasonable and which go really too far. And we need some sense of direction. We need to eliminate some paths if we are to pursue any with any seriousness. So my verdict on this inquest you can take, at least at this stage, to be no more than pragmatic opinion: it is that life originated on this Earth some 3 or 4 billion years ago from natural causes. That is a very conventional verdict nowadays, but the singular aspect of this case, that apparent gap at the start of evolution, will nevertheless lead to a view of the origin of life that is not at all conventional.

> 'Of course, if Dr. Mortimer's surmise should be correct, and we are dealing with forces outside the ordinary laws of Nature, there is an end of our investigation. But we are bound to exhaust all other hypotheses before falling back upon this one.'

2

Messages, messages

'What was the starting-point of this chain of events? There lay the end of this tangled line.'

There is a line that connects us to our ultimate ancestors – some not-yet-alive organisms that inhabited the primitive Earth. No doubt it is a somewhat tangled line: but what *sort* of a thing is it? What is the nature of the connection within a succession of organisms?

Every organism has in it a store of what is called **genetic information**. This is a set of instructions about how the rest of the organism, its **phenotype**, is to be made and maintained. I will refer to an organism's genetic information store as its **Library**. Man's Library, for example, consists of a set of construction and service manuals that run to the equivalent of about a million book-pages altogether. Simpler organisms, such as bacteria, make do with much less information in their Libraries. But even the thousand pages or so needed for a bacterium is still quite a weighty manual.

The pages in these figurative books are closely printed in a script that uses just four symbols. You can imagine these as, say, the letters *a* to *d* filling line after line of page after page with very little apparent rhyme or reason in the order of the symbols. Of course the lack of obvious order allows the possibility that a symbol sequence might be carrying messages of some sort. Although there are suspicions that some Libraries, such as man's, could often do with some crisp editing, there is no doubt that many if not most of the letter sequences do indeed hold messages of some sort. Indeed many such messages have been decoded.

An organism that is big enough to be visible to the naked eye is made up of a large number of compartments, or cells – usually of a variety of sorts with different functions. The materials of our bodies, materials such as skin, bone, blood, nerve, etc., are each made from a few sorts of characteristic cells.

Where is the Library in such a multicellular organism?

The answer is everywhere. With a few exceptions every cell in a multicellular organism has a complete set of all the books in the Library. As such an organism grows its cells multiply and in the process the complete central Library gets copied again and again.

Indeed the books in the Library can actually be made visible under the microscope just at the moment that cells are dividing and arranging that each of the two new cells will have a complete Library. Just before the cell divides pairs of stubby cord-like structures appear, seemingly clipped together at a point part of the way down their length. Then, when the cell divides, one cord of each pair goes to each of the two new cells. Each of these cords is an instruction manual, one of the enormous books in the Library; and the pairs are two copies of the same manual. Evidently this is all part of a system for an equal share-out.

The human Library has 46 of these cord-like books in it. They are called chromosomes. They are not all of the same size, but an average one has the equivalent of about 20 000 pages.

Of course chromosomes are not actually books with pages in them. A somewhat closer analogy for a chromosome would be a closely printed paper tape with the four kinds of letters in one long sequence. If you were actually to type out a tape equivalent to that in a typical human chromosome it would stretch for some 150 kilometres. It would be a long book whichever way you look at it. (Imagine trying to read a book like this on a windy day...)

Although chromosomes may appear as elongated objects the actual message tapes that they contain are much, much longer than the objects seen. The incredibly thin tape in a chromosome is coiled and supercoiled with incredible neatness. It would have to be neat because objects that are as light as these message tapes are continually being violently buffeted about by the molecules around them. (For the tiny components in cells it is always 'a windy day'.)

We humans are ENORMOUS animals. We have several million million cells in us with nearly as many copies, then, of our entire Library currently in print. Each cell equipped with so much information, has a certain autonomy. The cell is a particularly important level in the organisation of large organisms. It is at a level somewhat analogous to the level of an individual in society: a multicellular organism can be thought of as a tightly knit community of cells. We have only a vague idea as to how such a multicellular organisation can be built up and maintained. But at least we can see that the messages that must pass between cells to maintain their collaboration can be fairly simple in principle – messages of the kind 'refer to page so and so and do what it says'.

Fortunately we can forget about the problems of how cells communicate with each other when we are thinking about the origin of life. The important idea here is the autonomous nature of cells. Indeed most organisms on the Earth today are single cells. They can only be seen under the microscope. What we tend to think of as 'life on the Earth' – those organisms that are obvious and visible to us – is a comparatively recent innovation. Although, as already pointed out, there is good evidence for single-cell organisms having been on the Earth 2800 or more million years ago, it seems that it was only about 700 million years ago that well-organised multicellular organisms put in an appearance.

Even among single-cell organisms some are more complex than others. The simplest kinds of free-living forms that we know much about are bacteria. It is natural to wonder whether the very first organisms were not, perhaps, something like our modern bacteria. After all, we can see a general trend in evolution towards more complex creatures. It is sensible to take a hard look at the simplest organisms that we know of. Perhaps then we will be able to define that gap in our understanding that we call 'the problem of the origin of life'.

A favourite creature to talk about is called *Escherichia coli*. This is not the simplest bacterium, but an amazing amount is known about it. J. D. Watson has estimated that we perhaps know as much as a third of all the chemical reactions that are going on in *E. coli* – and that is a lot as you will see.

E. coli is a normal inhabitant of the human gut, but is quite capable of living on its own, given suitable nutrients. Only some bacteria are parasites: in this respect they differ from those still simpler 'half-organisms', the viruses, which have to be parasites and so could not have been a first form of life.

'Simple' is a relative term. Even viruses are not that simple, and by any absolute standards, *E. coli* is not simple at all.

True, *E. coli* is small by our standards – a rod a thousandth of a millimetre across and about twice as long. It is certainly not ENORMOUS. But it is enormous in its way. It is still vastly bigger than the components out of which it is made.

It is an indication of the sheer complexity of *E. coli* that its Library runs to a thousand page-equivalents. A better analogy would be a closely typed loop of paper tape: it would be about 10 kilometres long.

If we were to say that we now understand how organisms work, this would not mean that we understand the details of the way in which the masses of information in Libraries unburden themselves into fully working organisms. *E. coli*'s book is only partly read and understood, never mind the books in *our* Libraries.

No, the more complete understanding is at a more general level. We understand in principle how it is that a machine could reproduce itself in the kind of way that organisms can be seen to. We understand what such a machine has to be like. It turns out, for example, that such a machine, however it is made, whatever it is made of, has to have *something like* a message tape in it.

Think about it. How can characteristics in parents re-appear in offspring? How could such a thing ever happen – and then go on and on happening for millions of years?

Billy has inherited his father's eyebrows. What does that mean? Father's contribution to the making of Billy was a single sperm – and sperms do not have eyebrows. So what was it that Billy inherited?

There are two distinctions to be made here.

First we have to distinguish between characteristics and determinants of characteristics. Gregor Mendel realised in the 1860s that what must be passed on between generations of people or cats or pea-plants cannot be actual characteristics (tallness, eyebrow shape, flower colour, or whatever) but entities that somehow cause such characteristics to develop as the organism grows from its initial 'seed'. These entities were to be called **genes**.

The second crucial distinction to be made is between the inheritance of goods and the inheritance of information. Billy did not inherit his father's eyebrows in the kind of way that he may one day inherit his father's gold watch; but rather as he might inherit, say, the secret of how to make that special toffee that was the backbone of the family's confectionery business. In biology both goods and messages are passed on from one generation to the next. But it is the messages that are much the most important inheritance: only they can persist over millions and millions of years.

This distinction between goods and information is a case of the ancient distinction between *substance* and *form*. While a message may have to be written in some material substance, the message is not to be identified with that substance. The message as such is form. As such it can be reproduced again and again, amplified in principle indefinitely. Through copies of copies of copies...a message may be retained although none of the original material that held it persists.

Forms that can reproduce can be extraordinarily persistent; more stable, in a way, than substances. The complex abstraction that we call Beethoven's Fifth Symphony would not be easily destroyed. You would be sceptical of a newspaper headline that ran 'Fire destroys third movement of Beethoven's Fifth' or 'Opening bars of famed symphony stolen'. Your scepticism would be based on the knowledge that a symphony is not actually a thing; that there are many scores of this symphony in existence (i.e. messages about how to make a performance); and that these scores could easily be reproduced if and when wanted.

This way of persisting, by continually making copies, is certainly part of how organisms succeed. And the reproduction of organisms explains how it is possible for them to have such an extraordinarily complicated way of being. For any other kind of entity it would be hopeless to depend for survival on an intricate interdependence of complex components. Sooner or later something would go wrong and that would be the end of it. So long as a form is uniquely tied to a particular piece of substance, it is vulnerable to accident. (The Mona Lisa really could be destroyed by fire.) But for reproducing beings that caveat does not apply. Reproducing beings can be as complicated as they like. The question is whether their complication tends to improve the effectiveness of their reproduction; that is the only question. How the complexity in organisms arose is another matter; although here too we can begin to see how it might have happened – through natural selection, a process that applies uniquely to reproducing beings.

And we can begin to see how it must be that organisms reproduce. They reproduce through copying the messages that specify them – those very messages that are passed on between generations.

Now it is true that over the shorter term messages are not the only inheritance. There must also be goods, if only the actual books or tapes that hold the messages. Indeed much more than that is needed. The tapes must be read and acted on: a certain amount of automatic equipment will be needed to do this. You can imagine those kinds of machines in automatic factories that carry out instructions fed to them on a magnetic tape. Such machines convert a message into a specific activity. Hence everything can be made by following sufficiently voluminous instructions. Among other things, of course, we have to imagine that these automatic manufacturing machines are able to hammer together brand new automatic manufacturing machines... Then at least one of these machines has to be handed on with the messages to the next generation.

One can see, indeed, that when a cell divides more is divided out than just the books of instructions: material over and above the chromosome material is included in each of the two new packages. It is clear that this additional material must contain prefabricated reading and manufacturing equipment.

But the supremacy of the messages remains. Everything in the cell, including all that automatic manufacturing equipment, must be written about somewhere in the Library. If some of these messages happen to have to be read and acted on before a new cell is formed, that is a matter of timing that does not affect the long-term outcome. In the long term, after many generations, all that persist are the messages. Every actual thing, every particular collection of atoms, every particular piece of equipment, every particular water molecule, even every piece of every message tape, will eventually be destroyed or mislaid. Only the messages will survive, the messages themselves: because they are forms, and forms of a particular sort. They can be copies of copies of copies…

So please respect the humble bacterium that is playing this game. It can reproduce, it can evolve. *E. coli* must have some sort of long-term memory about how to make itself that can outlast its substance. That means that an *E. coli* must be an automatic factory containing something analogous to control tapes and automatic manufacturing equipment. And that is only part of it. All the equipment must be contained, organised, fed. Pieces for it to work on, energy to drive it, must be provided by the *E. coli* cell. Apart from the manufacturing machinery that can *follow* instructions, there has also to be another kind of machinery that instead *reprints* them – something analogous to a Xerox machine or a tape copier. All these things have to be contrived through the manufacturing machinery duly instructed by appropriate bits of the Library tape.

It may seem hardly surprising that no one has ever actually made a self-reproducing machine, even though Von Neumann laid down the design principles more than 40 years ago. You can imagine a clanking robot moving around a stock-room of raw components (wire, metal plates, blank tapes and so on) choosing the pieces to make another robot like itself. You can show that there is nothing logically impossible about such an idea: that tomorrow morning there could be *two* clanking robots in the stock-room…(I leave it as a reader's home project to make the detailed engineering drawings.)

There is nothing clanking about *E. coli*; yet it is such a robot, and it can operate in a stock-room that is furnished with only the simplest raw components. Is it any wonder that *E. coli*'s message tape is so long? (If you remember the paper equivalent would be about 10 kilometres long.) Is it any wonder that no free-living organisms have been discovered with message tapes below '2 kilometres'?

Is it any wonder that Von Neumann himself, and many others, have found the origin of life to be utterly perplexing?

Consider:

(i) Evolution through natural selection depends on there being a modifiable hereditary memory – forms of that special kind that survive through making copies of copies…, forms which can also be accidentally modified to produce modified effects. It is only effects produced like that that can have a long-term future. There can be no accumulation of appropriate accidents, no kind of progress, without the means to remember.

(ii) Successions of machines that can remember like this, i.e. organisms, seem to be necessarily very complicated. Even man the engineer has never contrived such things. How could Nature have done so before its only engineer, natural selection, had had the means to operate?

We are faced with an if-then-either-or.

If life really did arise on the Earth 'through natural causes' *then* it must be that

either there does not, after all, have to be a long-term hereditary memory for evolution,

or organisms do not, after all, have to be particularly complex.

As this is a detective story I will not say yet which way the argument will go – whether that 'if' will survive and, if it does, whether the 'either' or the 'or' will turn out to be the case.

> 'My head is in a whirl', I remarked; 'the more one thinks of it the more mysterious it grows.'

3

Build your own *E. coli*

I was relieved by this sudden descent from the general to the particular.

Organisms are, in places, built to an atomic precision – with their nuts and bolts as **atoms**. So pick your strongest figurative magnifying glass and let us see if we can get some idea of the principles of construction and mode of working of these most intricate of all mechanisms.

There are about a hundred chemical elements, that is to say about a hundred substantially different kinds of atoms. Rather more than a quarter of these are used in organisms. Six are pre-eminent:

> Carbon atoms
> Hydrogen atoms
> Oxygen atoms
> Nitrogen atoms
> Phosphorus atoms
> Sulphur atoms

These kinds are particularly good at forming **molecules**. A molecule is a more or less durable association of atoms. A molecule may contain any number of atoms – two or two million. What is important is that the atoms in a molecule are joined up in a particular way. Molecules are not just clumps of atoms: far from it, there are many big molecules in organisms which can best be described as *machines*.

Even atoms themselves are quite complicated things, with complicated inner structures. They are still rather mysterious, although we can say that the different kinds of atoms have different numbers of three more fundamental entities in them: electrons, protons and neutrons.

Except in creating the diversity of kinds of atoms, the deep subatomic levels of structure are not important for our purposes. Within organisms it is quite abnormal for this deep structure to change – for one kind of atom to change into another.

On the other hand there is a more superficial level of structure within atoms that cannot be ignored. One of the types of atom components – those strange light-weight particles of electricity, the electrons – can often attach and detach from atoms and molecules fairly easily. This gives rise to **ions**. These are atoms, or groups of atoms, that either have 'negative' electric charges because they have acquired one or more extra electrons; or 'positive' electric charges because they are one or more electrons short. Chlorine, for example, almost invariably occurs in cells as the negative ion (Cl^-), while sodium atoms are there with an electron missing (Na^+). (This 'positive' and 'negative' convention for describing the two kinds of charges may seem a little strange: the key idea is that charges of the same kind repel each other across space while charges of the opposite kind attract each other.)

Electrons are crucially involved in making those bonds between atoms – **covalent bonds** – that allow durable molecules to be built up. Electrons, pairs of them, are the rivets, and there is a continual shifting about of pairs of electrons as molecules are built up or taken apart, or as their structures are re-arranged.

For our immediate purposes we can be much more mundane. We can use as a model for an atom a bead of a particular colour with (for our purposes) a characteristic number of press-studs on it. (This number has to be modified for ions, but no matter.) For example, you can think of that smallest kind of atom, the hydrogen atom, as a small bead with a single press-stud. Two of these H-beads could be clicked together to make a model of a hydrogen molecule:

although this is usually represented on paper as:

the little stick being the covalent bond. Oxygen and sulphur atoms, on the other hand, would have two press-studs on each; nitrogen atoms would have three, carbon atoms four, phosphorus atoms five.

There are other rules that limit the ways in which atoms can be joined together to form covalent bonds, but even so there are vast numbers of ways of putting together atoms into molecules, particularly **organic molecules** – which have carbon atoms in them. Carbon atoms are excellent building units: not only do they have four

press-studs each, but they are particularly good at forming chains
and rings.

To give you an idea of the potential of this construction kit, here
are two ways of joining up nine carbon atoms together with 20
hydrogen atoms:

These are but two of 35 ways in which these 29 atoms can be joined
together. Each of these ways is a different pattern of connections, a
different way of joining the atoms so that every hydrogen is attached
by only one bond, while ever carbon has four links to adjacent atoms.
Each such way is a different molecule, a different substance. They
are distinguished from each other as one tune is distinguished from
another, being a different distinct way of arranging simple elements.

As the number of atoms in such molecules increases, the number
of ways in which these atoms can be arranged increases dramatically.
For example, there are over 60 million *million* ways of arranging just
40 carbon atoms plus 82 hydrogen atoms. In chemists' terminology
there are over 60 million million compounds of formula $C_{40}H_{82}$. Each
of these materials, with its particular kind of molecule, would
be – is – a unique material. (Examples of organic molecules containing also oxygen, nitrogen and phosphorus atoms are given in
appendix 1.)

The set of all the organic molecules that might be made is absurdly
large. It is like the set of all possible patterns defined by the rules of
western music, or the set of all possible chess games. Despite the rules,
there is still a universe of possibilities to be explored. But such a
universe only becomes accessible to exploration at a certain level of
expertise. It must be possible to discriminate between, and make,
particular arrangements. You can only compose music when, at the
very least, you can choose and arrange particular notes; you can only
play chess when you can select and move one piece at a time. To make
organic molecules, to explore that huge universe that organisms
explore, an atomic precision is required. Generally speaking it matters
which atoms are where. Most of the large molecules in organisms

are specified with that precision, and for the most critical functions, in printing and reading message tapes for example, they have to be.

So if you were to try to build a model of E. *coli* from bags of differently coloured beads, you would have to be very careful to stick the beads together in the right way.

Should you decide to tackle this project it is only fair to warn you that you will need a fair amount of space (say the nave of a cathedral) as well as time and money.

Assuming that you have decided to go ahead, you must first find a bead merchant who can offer you press-studded varieties at the keenest terms. If you can get them for 1p each so much the better, because even then the number that you will need will cost you about £2 billion. Perhaps you will be able to get a reduction on the water molecules which would be about 70% of your purchase – and I would advise you to get these molecules (H–O–H) pre-made.

Use beads that are as small as possible. They should be less than a centimetre across if you want your finished model to fit inside, say, Salisbury cathedral. You would have to think of some way of keeping this rather large sausage in shape. And the floor may need strengthening as your E. *coli* will weigh some tens of thousands of tonnes.

You will need plenty of well-trained staff. If each worker can click together beads at a rate of one every 5 seconds working 8 hours a day, 5 days a week, then a staff of a thousand should finish the job within 35 years. (I am assuming that you did not have to make the water molecules.)

Perhaps you will be discouraged from embarking on this project by the thought that a real E. *coli* does the equivalent of all this work in about half an hour: an equivalent amount of sheer manipulation must take place in the synthesis of all the molecules needed to make two cells out of one. It is true that E. *coli* does not use bare atoms as feedstock. Often it may pick up molecules, such as amino acids, that can then be used directly for making larger molecules. But, even then, there is a lot of manipulation: this food molecule has to be recognised and deliberately carried in by machinery in the skin of the bacterium. And more often than not the molecule taken in has to be largely taken apart to provide components that are then used to make the molecules that are really needed. Even just forming one covalent bond can involve several distinct operations – as, for example, when amino acids are joined up to make protein molecules. It is hard work whichever way you look at it.

Another admittedly discouraging thought is that even if you have

persuaded the bishop to fill the nave of his cathedral with your
magnificent model, it will not be generally appreciated. People will
say that it is just a bag of beads.

And they will have a point. Unlike *E. coli* itself your model would
not *do* anything. It would just lie there, a big sagging bag.

What would you have to do, then, to make a working model of
E. coli?

Part of the secret of *E. coli*'s working lies in the forces that operate
between molecules. Somehow your model would have to incorporate
these weaker **secondary forces** as well as the strong covalent bonds
that keep molecules intact. It is largely through such secondary forces
that molecules in organisms interact with each other. Molecule A,
for example, may associate with molecule B because A and B fit with
each other. You could think of A as like a hand, B as like a glove.
Even cleverer, the B-glove may be not quite the correct shape for A
so that there would be a better fit if A were distorted, if its internal
covalent bonds were strained a bit. Then some of the press-studs in
A may spring open to be re-closed in another way.

That is how **enzymes** work (at least it is part of the way). Enzymes
are proteins – molecules large anough to get a grip on other molecules
in their surroundings and transform them by re-arranging covalent
bonds. Enzymes are the main class of specialist machine-tools in
organisms.

Goodness knows how you would mimic these between-molecules
forces in your model (with little magnets perhaps?). It would
certainly add to the expense. Just suppose, all the same, that
somehow you had managed this. (You had even eliminated the effect
of gravity by putting the whole thing in orbit round the Earth.) Even
then you would still not have a working model: the forces between
atoms and molecules are not enough to bring about a continuing
activity. You would also have to *shake* your enormous bag (in orbit).
And you would have to keep shaking it with a carefully adjusted
violence. The violence of the shaking would have to be enough for
the bead molecules to jostle haphazardly from one place to another
in the bag. It would have to be enough, that is, to overcome, quite
often, those weaker secondary forces between molecules. On the
other hand the shaking should not be so intense that the covalent
bonds within the molecules break haphazardly.

Supposing that you have overcome these further technical problems, you would at last begin to see things happening. Molecules
would now have a chance to find each other. The 2000-odd kinds

of enzymes in your *E. coli* would have a chance, now, of bumping into the molecules that they were designed to transform. You would even see enzymes working.

I am afraid that your model would still not really work unless, too, your shaking machine had a suitably controlled *kind* of shaking. It must give a range of shakes from little ones to big ones in an appropriate proportion, and these different levels of jostling would somehow have to be evenly distributed throughout all the molecules in your model. I doubt if you would ever get the shaking quite right: I doubt if you could ever mimic the **heat agitation** of real molecules.

You see the missing factor, that made that bag of beads in the cathedral seem so dead, is the perpetual motion of atoms and molecules. They move, they spin, they vibrate, they jostle – and they need no power to do it.

The perpetual motion of atoms was predicted by Greek philosophers, but only really confirmed in the nineteenth century. It is quite as important for the understanding of chemistry as the atoms themselves. Molecules move too, although the bigger they are the slower they go. Indeed all objects are in a state of heat agitation being buffeted by the molecules around them, partaking in the general inescapable molecular motion. But if an object is big enough to be visible its net heat motion will be too tiny to see, although the violence of its inner vibrations, the vibrations of its atoms and molecules, can be assessed by touch, by how hot the object feels.

Notice that all this talk about atoms and molecules, the forces between them and their motions, is physics and chemistry. There is nothing special about the molecules in *E. coli* that they move so frantically. They are frantic because they are very small objects, and because it is not too cold. *This* piece of the magic, anyway, is not peculiar to life. What differentiates *E. coli* from just any speck of matter is in the detailed behaviour of its molecules. In *E. coli* they seem to have a purpose in their frenzy. That seeming purpose is to be found (and in the end found only) in the controlling messages. One might be tempted to see in the moving about of molecules in *E. coli* its most life-like feature – even the stamp of life. But really it is the message rather than the motion that is the hallmark.

'...it becomes not simpler but stranger...'

4

The inner machinery

'It is a capital mistake to theorize in advance of the facts.'

This is perhaps the most technical chapter in the book (although it is not *that* bad). Some readers may want just to skim it (or skip all but this page if they must), taking on trust the main burden of the argument that it presents – that the workings of all life on the Earth are seen to be fabulously complex and sophisticated on the molecular scale. Present-day organisms are manifestly pieces of 'high technology', and what is more seem to be necessarily so.

Get back to the tapes

What are the tapes made of that carry the genetic messages? What is the **genetic material**?

It is called DNA. Actually a piece of it is more like a long chain than a printed tape. The DNA chain has in it four kinds of links. Each of these links is quite a complicated object containing more than thirty atoms – of carbon, hydrogen, oxygen, nitrogen and phosphorus – joined up in a particular way. The links are **nucleotides**. (Their detailed structures are given in appendix 1.)

Here are jigsaw-piece analogies for the four DNA nucleotides:

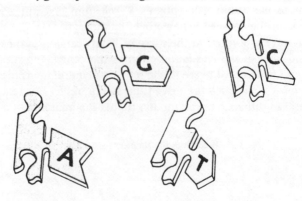

Notice that there is a common connector piece in these units (on the left-hand side as I have drawn them) which would allow you in principle to make a chain with no restriction on the sequence of the four kinds of letter pieces. This is clearly a suitable arrangement for making chains with messages written in them.

Nucleotide units do not join together on their own, even with the help of the heat agitation. To get them to join up they have to be **primed** – 'wound-up', so that they have more energy in them. Their structures have to be modified, to be set like mousetraps. Then they are able to snap together into chains. An extra piece is attached to each of the nucleotide units in order to prime them, these pieces breaking off as the chain forms. (This is described in more detail in appendix 1.)

Given supplies of the four kinds of links, duly primed, the next question is how the sequence of linking is chosen. We know there has to be some sort of copying process, so we can put the question like this: How is the sequence in some newly forming chain determined by the sequence in some other chain already there?

I can remember the excitement in the early fifties when it was discovered that DNA had a *double* chain, and that a sequence in one of these chains clearly determined the sequence in the other. It was almost as though these great long molecules had been caught in the act of printing off copies of themselves. It was rather like finding a whole lot of photographic prints stapled together with their negatives. The technique of reproduction seemed a give-away.

In terms of our jigsaw model this is what the structure was found to be like:

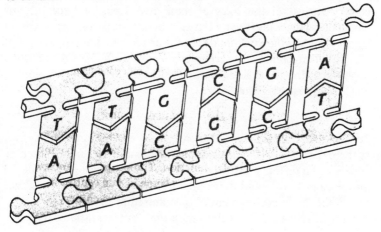

You can see that the letter sequence in one of the chains in a DNA molecule is complementary to the letter sequence in the other. Wherever A is the letter on either strand then the opposite letter, on the other strand, is T. Similarly G on one strand lies exactly opposite C on the other. If you look carefully at the jigsaw model you will see that a large letter has to pair with a small one, while at the same time a 'plug' must go with a 'socket'. (The real molecule is not flat like this model, but twisted into a double helix, like an old-fashioned lamp flex, with the letter pieces pushed up against each other on the inside.)

Now you can imagine the double DNA message replicating. Suppose the strands begin to come apart at one end, or unwind somewhere in the middle. Either way the exposed single chains now attract primed nucleotide units in a complementary fashion. These units link together on each of the single chains turning them into new stretches of double chain. Unwinding and chain-making continues, and the final outcome is a pair of identical double chains in place of the original one. (The reality is much more complicated, but this gives the general idea.)

The forces that are responsible for choosing the new units, so that they match up, are secondary forces of the kind that we discussed briefly in the last chapter. While the covalent bonds that hold the units together through their connector pieces are formed (more or less) once and for all, the forces between the letter pieces – the 'plug–socket' pairing forces – are much less emphatic. They are more exploratory. The units come and go and come and go many times before an appropriate pairing between letter pieces is accidentally made. This is very typical of the roles of these two kinds of forces in our biochemical machine: the tentative, exploratory, secondary forces set up a situation which culminates in the decisive making (or breaking) of a covalent bond. The carpenter must first choose and carefully align the appropriate pieces of wood before the decisive act of pinning or glueing them together.

What do the messages mean?

To a first approximation the messages in E. *coli* mean **proteins** – two or three thousand different kinds of protein molecule, each a machine with a more or less particular function. The purpose of most of the messages in E. *coli* (and that goes for some of our central messages too) is the direct specification of this molecular machinery.

A protein molecule contains a message – but a translated message, a translation of some passage in some book somewhere in the Library of the organism that made it.

The protein language is more interesting than the DNA language in two respects. First, there are twenty, not four, letters. Second, the letters are a far more varied set. Instead of looking all much the same, there are small ones and big ones, long ones and fat ones, floppy ones and stiff ones, sticky ones and smooth ones, ones with negative electric charges and ones with positive charges...

The protein units are called **amino acids**. These are smaller than nucleotides, the smallest one having only ten atoms in it, the largest twenty-seven. (Some examples are given in appendix 1.) Again, as with the nucleotides, part of each unit is a common connector piece. Through these, the twenty kinds of letters – the varied side pieces – can be put into specific sequences. And again these amino acid units have to be primed before they can be joined together. A typical protein is a particular string of between a hundred and a thousand amino acids.

You might think that a protein molecule would be like an enormous (open) charm bracelet; or like a washing-line with hundreds of items (of twenty different sorts) hanging from it. What *could* such a washing-line message mean?

Very often it means 'Fold up like this'. The units being smaller, a string of amino acids is more compact than a DNA string. This, and the variety of its letters, encourages particularly interesting and complicated forms of folding. The cohesion of the folding is helped by the chain of connector units being rather sticky in itself; but the nature of the folding is decided by the arrangement of the letters. All their odd shapes try to fit together under the pull of the secondary forces; and the groups that have a strong attraction for water – especially those that hold an electric charge – try to get to the outside. It is a very complicated calculation how best to fold up so as to give in as far as possible to the great variety of secondary forces that are at work – to get everything neatly packed together and yet leaving a little elbow room for the heat agitation. We have yet to teach our best computers to make such calculations. But the squirming message tape quickly gets the answer. In almost no time that message that had been translated from a central Library, from DNA language to protein language, has transformed into a piece of machinery that works. At last the message says something in the most direct way that you can imagine: it becomes something.

Very often the protein message becomes an enzyme: one of

thousands of machine-tools that together make the molecules on which the whole enterprise runs. Enzymes make the nucleotide units for DNA, for example, and also amino acids. Molecules like these may need ten, twenty, thirty separate steps in their making – and as many different enzymes, each specially designed to carry out one step, one re-arrangement of covalent bonds. And there are many other kinds of molecules that are made.

Among such components there are lipids, fat-like molecules, needed for the cell membrane. In *E. coli* this is the inner of its two skins. The cell membrane contains in addition many proteins that organise the lipid molecules and create selective channels or pores. There are also protein machines in the membranes of cells that actively pump selected materials out and in.

One of these pumps in the cell membrane of *E. coli* is a hydrogen ion pump that acts rather like a battery charger, maintaining a kind of voltage between the inside and outside of the cell. The release of this voltage drives other pumps. It also drives turbines that rotate propellers by means of which *E. coli* swims about.

Protein really is the stuff of life. The parts of cells that are not made *of* proteins are at least made *by* proteins. Even the DNA message tapes have their component units manufactured and joined together with the assistance of protein enzymes.

Mindless translation

How does the translation take place between the austere DNA language of the central Libraries and the protein language, the language of action?

How can a message in a language that has only four kinds of letters be translated into a message in a language that has twenty kinds of letters? There are several possible solutions to this formal problem. In fact organisms employ one of the simplest: the DNA letters are (in effect) read in threes. That immediately gives 64 possible 'words': AAA, AAG, etc., etc. Every such 'word' corresponds to a letter in the protein language (or to a full stop). The 64 possible 'words' are far more than are needed, and it is usually the case that two or more different DNA 'words' correspond to the same amino acid.

The immediate problem is not formal, but practical: How in fact does the translation get done automatically, mindlessly?

The brief answer to this question is with off-prints, a set of adaptors, lots of big enzymes, and a huge machine. Let me explain.

The off-prints are working copies of small parts of DNA Libraries:

they are tapes copied off one strand of a DNA molecule that has been partly unwound for the purpose. The off-print itself is a single strand of RNA. RNA is the other kind of nucleic acid: it has a similar structure to DNA (see appendix 1).

The adaptors are also made of RNA although they have a quite different immediate function. They are not message tapes but neat little pieces of machinery. Each is a single strand of RNA, about 80 units long, that is twisted upon itself in a specific manner. The pattern of twisting is determined by the sequence of the letters, many of which pair with each other. The result in each case is a rather fancy kind of three-pin plug. The pins consist of an exposed unpaired triplet of RNA letters. The different kinds of these adaptors all have a similar shape, but they have different exposed triplets that will plug into different complementary triplets on message tapes.

The big enzymes that I talked about are each able to select an appropriate amino acid from their surroundings, as well as one kind of adaptor, and join them together. For example, an adaptor with CCC as the exposed triplet would only be loaded with the amino acid glycine. This is because one of the words for glycine in nucleic acid language is GGG, and a CCC adaptor would, in suitable circumstances, stick to that word.

The huge machine is called a ribosome – and it provides 'suitable circumstances'. It is built out of both RNA and protein molecules. The ribosomes in *E. coli* each have about 270 000 atoms in them (and there are about 30 000 of them at work in one *E. coli*).

It is the ribosomes that actually make proteins by organising, both in space and time, the interactions between RNA message tapes and suitably loaded adaptors.

To operate, a ribosome attaches itself to a message tape and runs along its length translating the message in the process into a growing protein chain. The chain falls off, the product is complete, when the end of the message is reached.

Suppose that you were to examine a ribosome part-way along a message tape, say just after it had linked on its fiftieth amino acid. You would find, then, that the 50-long chain was attached through the amino acid that had just been added. That is the way a protein chain grows, like a blade of grass, from its base. Looking more closely you would find that the whole chain was attached to the adaptor for that fiftieth amino acid, and that the other end of this adaptor was plugged into a corresponding word on the message tape – adaptor and tape being held within the ribosome. Also within the ribosome would be the next word on the tape. The ribosome and this part of the tape together would be creating an empty three-pin socket. Suppose that this next part of the tape read GGG: then an adaptor with CCC pins would fit this socket, an adaptor carrying glycine.

Of course there is no operator there to reach for a glycine-loaded CCC adaptor and push it into place. There is only the heat agitation of the molecules to allow a random exploration, and the extra firmness of a correct fit as the indication of success. One by one various adaptors, as well as many other molecules, collide with the empty socket. Eventually a loaded CCC adaptor arrives, happens to collide the right way round, is accepted and clicks into place. This allows the next most crucial step. It seems very precarious: a covalent bond is broken while another one is made so that the whole chain on the adaptor for the fiftieth amino acid is transferred to the amino acid on the adaptor next-door. The adaptor for the fiftieth amino acid is now empty and is rejected. The ribosome moves three RNA letters on to complete a cycle of operations. The situation now is rather as we found it, except that now there is a chain of 51 amino acids attached to an adaptor plugged into the message tape located in the ribosome...

Recapitulation: An essential complexity?

In this chapter, and in the last two chapters, I have been trying to give an outline of the central workings of organisms – all the organisms on the Earth as far as anyone knows. Right at the centre are the DNA messages, the only connections between life now and life a million or a billion years ago. Only these messages survive over the long term, because only these messages can persist through the making of copies of copies of copies...

So here is how the life that we know works. DNA makes DNA (given primed DNA nucleotides and enzymes); DNA makes RNA too (given primed RNA nucleotides and more enzymes); and then RNA – RNA messages, RNA adaptors, RNA in that huge machine – makes proteins (given amino acids, the means to prime them and still more enzymes). The proteins (especially enzymes) do everything else.

Too simple? Well, yes, it is a bit too simple a description of today's organisms: but it is also far, far too complicated as a description of a first organism. The worst bit is that much of the complexity seems to be necessary: if you are going to have a form of life whose replicable messages are written in nucleic acid (either DNA or RNA), and which operates via protein, you are surely going to be landed with a very complicated system. It is like, say, video-recording: if you want to do that sort of thing – use magnetic tape to record moving pictures – then you are going to have to pay for it: there will be no very easy way: video-recorders just are complicated machines. An *E. coli* just is a complicated machine too, and I think that *any* free-living nucleic-acid-based forms of life would have to be.

Take just part of our system – the automatic protein synthesiser. Any such machinery, however it is made, is surely going to be clever, complicated engineering; because it is a complicated and difficult job that has to be done.

Ask any organic chemist how long it takes to put together a small protein, say one with 100 amino acids in it. Or go and look up the recipe for such an operation as it is written out in scientific journals. You will find pages and pages of tightly written instructions, couched in terms that assume your expertise in handling laboratory apparatus and require you to use many rather specialised and well-purifed chemical reagents and solvents. And the result of following such instructions? If you are lucky a few thousandths of a gram of product from kilograms of starting materials.

Or go and read all the details and examine the engineering drawings for a laboratory machine that can build protein chains automatically. (If you want to buy one it will cost you more than a video-recorder.) You will be impressed by how clever such machines are – and not surprised that *E. coli*'s machine is clever too. It would have to be, wouldn't it?

Notice furthermore that the making of proteins in organisms is under instructions from replicable messages. This is no added extra feature that might have been dispensed with in earlier, simpler designs. It is essential to the whole idea. Protein or protein-like material made otherwise would not have been directly relevant because it would not have been subject to elaboration through natural selection: it would have been disconnected from the successions of messages that alone maintain the long-term continuity. Nothing evolves that is not somehow tied into the successions of messages.

Nor could the precision of manufacture have been much less if it was enzymes that were needed right away. A clumsy enzyme is a good bit worse than useless if it is continually transforming molecules the wrong way, or transforming the wrong molecules. (Enthusiastic incompetence is much worse here than sloth.) More and more molecules would be produced that had been wrongly put together, and these would include components for RNA adaptors, ribosomes, etc. – leading to further badly made enzymes and a rapid slide into chaos.

Nor does it take much for an enzyme to become incompetent. The whole technique of operation requires that the protein message folds up in a way that depends on the sequence of amino acid units. Even having only one mistake, one wrongly inserted amino acid, can wreck any chance of a correct folding; and more than a few mistakes are almost bound to.

To use a big folded molecule to make and break other molecules (like using magnetic tape to hold moving pictures) is a marvellous idea – if you have the technology.

Finally, and again casting back to chapter 2, it is not just the sheer size of even the smallest Libraries; it is not just that nucleotide units are rather complex in themselves, and rather difficult to join together (because Nature is on the side of keeping them apart); it is not just the need for enzymes, here, there and everywhere; it is not just that enzymes are of little use unless they have been made properly; it is not just that ribosomes are so very sophisticated – and look as though they would have to be to do their job; it is not just such questions relating to the particular kind of life that we are familiar with. There seems also to be a more fundamental difficulty. Any conceivable kind of organism would have to contain messages of some sort and equipment for reading and reprinting the messages: any conceivable organism would thus seem to have to be packed with machinery and as such need a miracle (or something) for the first of its kind to have appeared.

That's the problem.

'...the thing becomes more unintelligible than ever.'

5

A garden path?

Lestrade laughed indulgently. 'You have, no doubt, already formed your conclusions from the newspapers', he said. 'The case is as plain as a pikestaff, and the more one goes into it the plainer it becomes.'

Seeing things

Perception is usually based on very limited data, as conjurers and artists know. A few lines with a pencil, or a few patches of colour, may be enough vividly to represent an object. You can see the object in a mere sketch.

Even a real object is usually 'a mere sketch' – for all the data about it that you are likely to have taken in. 'It's a wooden chair of course', you say, after the most cursory glance. Actually your eyes have only picked up some of the light scattered from some of the surfaces; you have assumed four legs although only three are in view; you have failed to check whether the object is solid, never mind if it is really made of wood as you suppose. You hardly know anything about the chair. Yet there you go, jumping to conclusions. You even add 'of course'.

Yet (of course) it makes sense. All you need are cues most of the time. When half a dozen input signals have checked out as 'chair-like', you do not bother to get out the little drill or the weighing machine. If it looks like a chair enough, then it is a chair. You have learned from experience that a guess based on just a few data is usually right. But a perception remains a (preconscious) guess even when it presents itself to the consciousness as an obvious fact.

Sometimes, though, something goes wrong. A perception falls to pieces and a new one has to be made. A pair of lights in the distance are the data on which you base your perception of an approaching motor car. But the lights start to move in relation to each other in an unexpected way. The car suddenly becomes two motorcycles.

Such occasions (when we say to ourselves 'Wait a minute, what's happening – oh, yes, I see') are familiar enough: they serve to emphasise how strong is the desire to get data categorised, converted to a perception, understood – and as quickly as possible. That we have such a desire is hardly surprising: to jump to a (correct) conclusion may be a matter of life or death.

In science we see things too: we make guesses to accommodate the facts.

Science gives us a perception of the world far beyond our senses. It is not literally a perception, like the perception by you, just now, of that chair; but it is the same kind of thing in that limited data are used to create a coherent 'picture' of phenomena. A shared perception of the world is arrived at, by what is called 'the scientific community'. It is a many-brain perception.

If the guesswork of our individual perception can lead us astray – totally astray at a conjuring show – so too can the perceptions of science. The trouble is that however many facts may seem to confirm some preception of phenomena, we can never have all the facts; we can never have observed everything all the time and in every possible way. As in seeing that chair, we must arrive at our understanding on the basis of the most minute fraction of all the conceivable evidence: and any such limited collection of data can always be fitted to alternative general views.

For example, the perception that the sun rises every morning, crosses the sky, and sets in the evening, is the everyday common-sense perception. But there is an alternative general view, the perception of science, that the Earth rotates. This is an alternative way of interpreting the immediate facts. How can one ever know that there are not still better explanations that have never been thought of?

In spite of all this we develop a confidence in our scientific perceptions of the way the world works. It is similar to the confidence we have in our everyday perceptions and rests on a similar base. We are pretty sure the chair is real. If need be we can become more and more sure that it is not a deception or a hallucination by taking in new data – preferably of diverse kinds. We look at the chair from a new angle, we lift it up, kick it, sit on it. As the new data continue to check out as 'chair-like', vestiges of doubt are soon removed.

Scientific perceptions – concepts, insights – cannot literally be kicked to see if they are real; but we do something similar when we check out an idea. New data are sought from as wide an area as possible. Do the new pieces fit into the general picture?

Often it is difficult to be sure. If the picture is rather vague, or if the pieces are too soft and malleable, it may be possible to go on fitting new evidence for a time to a false picture – even if a feeling of unease grows. ('This can't be right', you may begin to say.) Your impression of something wrong may precede the insight as to what it is.

It is natural, when new scientific evidence does not fit an accepted picture, to see if one can bend the picture. Often enough that works. Indeed the picture may even be improved by the manipulation: it may become simpler or more general. Few things are more convincing, indeed, than when a piece of evidence that at first seemed to go against a theory turns out, with some small adjustment, to fit particularly well (the so-called 'exception that proves the rule').

On other occasions the perception collapses. It is not modifiable and has to be replaced by something quite different. 'There is something wrong' changes to 'This is wrong altogether'.

Anyone familiar with scientific research (or detective stories) will know what I mean here. You make a guess on the basis of a few bits of evidence; you see if the guess holds up with more evidence; when it doesn't you first try to modify the guess; when it still doesn't you try another guess, perhaps an altogether different one. That sort of thing goes on all the time in science: it is called trying to work out what's happening.

Rarer and more spectacular are the cases where a misguided insight comes right out into the open to become, for a time, a generally accepted doctrine. The most famous case in chemistry was the phlogiston theory of fire. This raged (the theory) throughout the eighteenth century. It was an attractive, common-sense idea: it said that when something was burned, a substance – phlogiston – was given off. It was the characteristic of inflammable materials that they contained this substance. When a piece of coal or wood or paper is reduced to ashes something has obviously gone away – the fire-stuff, phlogiston.

The idea was extended to metals. The rusting of iron was also a giving off of phlogiston, this same phlogiston that all metals contain. (This is why all metals are shiny, by the way, and look rather similar, while their rusts are much less uniform looking.) Living organisms too were seen to be rich in phlogiston and the life process a slow kind of burning.

It was a good theory, in its way, with a considerable coherence. That burning, rusting and respiration are closely related was a correct insight. Many of the great chemists of the eighteenth century believed in phlogiston.

The staying power of the phlogiston theory lay partly in the comprehensiveness of its error: it was almost exactly the opposite of what is the case. For 'phlogiston' read 'absence of oxygen', and you are not far off. Many of the connections were correct within the phlogiston scheme – except that they were the wrong way round.

The initial plausibility was doubtless another factor in the persistence of the phlogiston idea. It readily caught on and was difficult to displace. You had to get in close before you could see that there was anything wrong. It was necessary to weigh things, to recognise gases as substances, and so on. It was through the finicky details that the unease began to grow. Phlogiston gave a picture of the overall phenomena, but it failed to provide satisfactory explanations for more detailed effects. A new synthesis of the phenomena was required, a new key – oxygen.

How is a new synthesis arrived at?

The answer is through analysis. The old picture must be taken apart. This replaces a state of some understanding by a state of some bafflement. It goes against the grain: it is perception in reverse. It does not seem to be the right way to go.

A new insight often seems to occur to people when, for a time at least, the pressure to get on with the job has been relaxed. (Archimedes in his bath is the archetype.) Then the perception of a problem can be toyed with, disassembled into component data and ideas. What sets the mind off in this analytical direction is not the understanding of something, but a failure to understand.

It is characteristic of thoughtful people that they don't understand some things that to others are as plain as a pikestaff. Newton didn't understand gravity – which to everyone else was obvious. (Why is that apple moving towards the Earth?) Einstein didn't understand light. (What would happen if one rode on a light beam and looked in a mirror?) And of course Sherlock Holmes was always being puzzled by seemingly obvious or trivial things. Understanding is all very well, but not understanding can be much more interesting. Hence the concentration, so far in this book, on what is appallingly difficult about the problem of the origin of life.

There are many thoughtful and knowledgeable people, nowadays, who don't understand the origin of life. This is in spite of a 'big picture' provided by a theory known as 'chemical evolution'. Like the phlogiston theory, 'chemical evolution' looks good from a distance, and there is a common-sense about it. But, to my mind, like the phlogiston theory, it fails to carry through an initial promise: it fails at the more detailed explanations.

'*Chemical evolution*': *a modern phlogiston?*

According to the doctrine of chemical evolution, molecules of the sorts that we now find in organisms were made originally without organisms. These molecules (amino acids, nucleotides, lipids and such) were made as a result of chemical and physical processes operating on the early Earth. The molecules were then further organised (by chemical and physical processes) into the first beings able to evolve through natural selection. Thus 'chemical evolution' can be seen as a first phase in an evolution from atoms to man. Chemical evolution is not the same as biological evolution, but nevertheless the two kinds are connected, and similar in their progressions from simple to complex.

It is a grand vision that seems to me to be a mix of things that are true and things that are not. But let me pretend, for a page or two, that I am a wholehearted chemical evolutionist. What should be the drift of my argument?

I would start from the unity of biochemistry – the second of the three prime facts of the case (p. 5):

'Surely there is a deep significance in the observation that of the millions of millions of possible organic molecules, all life that has been discovered so far is based on a mere one or two hundred units – molecules of the size of amino acids or nucleotides that contain from 10 to 100 atoms. "The molecules of life" they have been called. Surely a life that is made so universally from these components must have been made originally from them? The Earth must have been the source of these molecular pieces: these molecules were either made by Earth processes or they were acquired in (e.g. meteorites) from space. It stands to reason, it is as plain as a pikestaff, that if a machine has to be made out of certain components, then the components have to be made first.'

I would then move to my next major point – that 'the molecules of life' are easy to make – and continue on these lines:

'Organic molecules could have been made under the influence of various forms of energy that would have been there on the early Earth – particularly ultraviolet sunlight and lightning – acting on constituents of the early atmosphere. Experiments have shown this. Amino acids and some other "molecules of life" form when sparks are passed through mixtures of gases simulating a primordial atmosphere. The best results are obtained here with atmospheres containing methane. But many other gas mixtures and sources of energy have proved effective. The main thing is that oxygen gas should be absent: but then it would have been absent on the primitive Earth, before there were plants to produce it.'

'Hydrogen cyanide is a small molecule containing one atom each of carbon, nitrogen and hydrogen. It can be made fairly easily (e.g. with sparks) in an atmosphere that has methane in it and nitrogen in some suitable form (e.g. nitrogen gas and a little ammonia). And hydrogen cyanide molecules can join together to make adenine, which is one of the nucleotide letters, as well as molecules related to the other letters. Amino acids can also be formed from cyanide.'

'Formaldehyde is another key molecule. Again it is very small, containing only one carbon, one oxygen and two hydrogen atoms. It can be formed in a number of ways – for example from ultraviolet sunlight on minerals in the presence of water and carbon dioxide. The wonderful thing about formaldehyde molecules is that they easily join together to give sugar molecules of all sorts. Several "molecules of life" are simple sugars. These contain carbon, oxygen and hydrogen atoms in the same proportion as in formaldehyde. Glucose is such a molecule. Often they have a ring of atoms, making a fairly rigid little unit that is useful for building purposes. (The connector pieces of nucleotides contain sugar units.)'

'Amino acids, sugars, nucleotide letters, could all have been formed on the primitive Earth. Indeed some amino acids seem to be quite easy to make, to judge from the way they keep on turning up all over the place. They can be found in some meteorites, for example: the very kinds of meteorites that have been least altered since the very origin of the solar system, before the Earth was born. And those small precursors of "molecules of life", cyanide and formaldehyde, are present in vast amounts throughout the Universe: in the huge spaces between the stars, in comets...'

'You can see it, can't you? A cosmic molecular preamble, a Universe itching to be alive.'

And if you ask me how the next stage happened, how the smallish 'molecules of life' came together to make the first reproducing evolving being, I will reply: 'With time, and more time, and the resource of oceans.' I will sweep my arms grandly about.

'Because, you see, in the absence of oxygen the oceans would have accumulated "the molecules of life". The oceans would have been vast bowls of nutritious soup. Chance could do the rest. Combinations of molecules came and went. Some combinations were more stable than others, forming coherent little droplets or clots in the soup. In some of these there were chemical reactions going on which had the effect of causing the droplet to absorb new material to itself, to grow. All the time breaking waves tended to disintegrate the droplets into smaller pieces that also absorbed new material to themselves... Eventually one such association of molecules – it would only have to be one – made it onto the ladder of Darwinian evolution: it could reproduce and pass on characteristics to offspring...'

'Of course the story is a bit vague on some of the details, but as a general view it seems sensible enough. Doesn't it?'

What is wrong with that sort of story?

What is wrong with the story that I have just been telling is that it hardly touches the real difficulties: the difficulties that I was piling up over the first four chapters.

I will grant that the path of chemical evolution seems sensible and in the right direction. There are a few obvious puddles to be avoided and some of the flagstones are a bit uneven, perhaps, but there is the promise of an easy walk up to the foothills of the mountain that we can see straight ahead of us. It is a promise that is unfulfilled. The trouble with this path is that it leads us toward, but it does not lead us to expect, a sudden near-vertical cliff-face. Suddenly in our thinking we are faced with the seemingly unequivocal need for a fully working machine of incredible complexity: a machine that has to be complex, it seems, not just to work well but to work at all. Is there cause to complain about this official tourist route to the mountain? Is it just a garden path that we have been led along – easy walking, but never getting anywhere?

I think it is. And I think we have been misled by what seem to be the two main clues: the unity of biochemistry and what is said to be the ease with which 'the molecules of life' can be made. If you take a quick look at these signposts they seem set straight towards our distant visible goal. But this straight route leads us to the cliff-face. Have we misread the signposts?

> 'That all fits into your hypothesis, Watson. But now we come on the nasty, angular, uncompromising bits which won't slip into their places.'

6

Look more closely at the signposts

'Holmes', I cried, 'this is impossible.'
'Admirable!' he said. 'A most illuminating remark. It *is* impossible as
I state it, and therefore I must in some respect have stated it wrong. Yet
you saw for yourself. Can you suggest any fallacy?'

What kind of unity?

Fortunately there is a good bit more to be read into the unity of
biochemistry than that all organisms on the Earth share a set of
molecules. There is plenty of small print to be scrutinised – and such
scrutiny quite alters first impressions. So let us move in with a hand
lens, and be fussy about four points in particular.

Point one. There is a SYSTEM common to all life on Earth, not just
a set of molecules. The unity of biochemistry applies, for example,
to the ribosome technique of protein synthesis; to the idea of using
proteins for catalysts, and of making membranes from proteins and
lipids. It applies too to more particular manufacturing procedures:
to the sequencing of operations in the making and breaking of
molecules. These 'central metabolic pathways' are extraordinarily
similar in all forms of life that we know of. (And their more detailed
structure is very revealing – but we will leave this further analysis
until the next chapter.)

Now it is quite clear that the universality of all this higher-order
organisation cannot be accounted for in terms of the pre-existence
of precisely this organisation on a lifeless Earth. I don't think that
anyone has suggested that the ribosome was picked out of a
'probiotic soup'. That being so it becomes correspondingly less clear
which, if any, of the component molecules might have been pre-
selected. At least some of the universality was an evolutionary
product. The thought arises: perhaps it all was.

But surely, you might say, you need the parts before you can make
a machine: 'the molecules of life' must have been there before life,
before the system could start to be built up. Is this not common-sense?

Common-sense? Well, perhaps it is. But it is a confusion. It is a false intuition based on what *we* think would have been a sensible procedure.

We may make a machine by first designing it, then drawing up a list of components that will be needed, then acquiring the components, and then building the machine. But that can never be the way that evolution works. It has no plan. It has no view of the finished system. It would not know in advance which pieces would be relevant. Even if amino acids (for example) had been in a 'probiotic soup', what use would they have been, long before their key use now (to make protein) had been hit on? *It is the whole machine that makes sense of its components.*

Point two. Subsystems are highly INTERLOCKED within the universal system. For example, proteins are needed to make catalysts, yet catalysts are needed to make proteins. Nucleic acids are needed to make proteins, yet proteins are needed to make nucleic acids. Proteins and lipids are needed to make membranes, yet membranes are needed to provide protection for all the chemical processes going on in a cell. It goes on and on. The manufacturing procedures for key small molecules are highly interdependent: again and again *this* has to be made before *that* can be made – but *that* had to be there already. The whole is presupposed by all the parts. The interlocking is tight and critical. *At the centre everything depends on everything.*

There are then four subsidiary points.

(*a*) It is no surprise that our central biochemical machinery is now so conservative: when everything depends on everything it is difficult for anything to be changed.

(*b*) Such a multiple interlocking of functions could only have been a product of evolution. The centre of our system then *became* fixed; but it could not have been fixed to begin with.

(*c*) This strong dependence of subsystems on each other is understandable as an evolutionary product in that it is typical of efficient pieces of machinery. A motor car, a clock, a television set, an oboe, a refrigerator, a tennis racquet...think of almost any sophisticated piece of engineering and you will find more or less diverse components, of little use by themselves, working in collaboration.

(*d*) Less clear is how a gradual step-by-step evolution can lead to a system in which everything depends on everything.

Chapter 8 will be a more detailed investigation of point two. The puzzling subsidiary point (*d*) will prove to be particularly helpful.

Point three. The common system is very, very COMPLEX – quite apart from the interlocking nature of that complexity. It was, then, the product of an extended evolution. *It is far, in evolutionary terms, from the original organisms.*

Point four. There are CONVENTIONS in the universal system, features that could easily have been otherwise. The exact choice of the amino acid alphabet, and the set of assignments of amino acid letters to nucleic acid words – the genetic code – are examples. A particularly clear case is in the universal choice of only 'left-handed' amino acids for making proteins, when, as far as one can see, 'right-handed' ones would have been just as good. Let me clarify this.

Molecules that are at all complex are usually not superposable on their mirror images. There is nothing particularly strange about this; it is true of most objects. Your right hand, for example, is a left hand in the mirror. It is only rather symmetrical objects that do not have 'right-handed' and 'left-handed' versions.

When two or more objects have to be fitted together in some way their 'handedness' begins to matter. If it is a left hand it must go with a left glove. If a nut has a right-hand screw, then so must its bolt.

In the same sort of way the socket on an enzyme will generally be fussy about the 'handedness' of a molecule that is to fit it. If the socket is 'left-handed' then only the 'left-handed' molecule will do. So there has to be this kind of discrimination in biochemistry, as in human engineering, when 'right-handed' and 'left-handed' objects are being dealt with. And it is perhaps not surprising that the amino acids for proteins should have a uniform 'handedness'. There could be a good reason for that, as there is good reason to stick to only one 'handedness' for nuts and bolts. But whether, in such cases, to choose left or right, that is pure convention. It could be decided by the toss of a coin.

It is one of the most singular features of the unity of biochemistry that this mere convention is universal. Where did such agreement come from? You see non-biological processes do not as a rule show any bias one way or the other, and it has proved particularly difficult to see any realistic way in which any of the constituents of a 'probiotic soup' would have had predominantly 'left-handed' or 'right-handed' molecules. It is thus particularly difficult to see this feature as having been imposed by initial conditions. Here again it

would seem that the convention, to be 'right-handed' or 'left-handed',
was arrived at concomitantly with the evolution of the whole
complex system. Looked at this way the problem of where the
'handedness' convention came from is all part of the same problem
as every other manifestation of the unity of biochemistry – and there
is a classical Darwinian explanation, then, that stares us in the face.

Descent from a common ancestor is a frequent explanation for
common features in different species of organism. All mammals have
hair because the single species from which cows, rabbits, whales,
people, bats, etc., evolved was hairy, and because hair had become
a fixed characteristic by the time this common ancestor appeared.

Sometimes one can be misled. The wings of the bird and the beetle
were discovered independently; they are not to be explained in terms
of common ancestry. So one has to be careful. But where a feature
is functionally not critical; where it appears to be a mere convention;
where it could very well have been otherwise – then one is on
stronger ground. If there are many shared characteristics of this sort
then one is on very strong ground. That there are many 'conventional' aspects of our central biochemistry allows us to assert with
some confidence what is a very widely held view that *the unity of
biochemistry arises because all organisms on Earth are descended from a
single common ancestor within which certain features had already been
fixed*.

Of course the fixed features referred to are precisely those features
that constitute the unity of biochemistry. We might surmise (from
point two) that these features became fixed through interlocking of
functions. In any case (from point three) this last common ancestor
of all life on the Earth was already highly evolved. Its fixed conventions were to be passed on eventually to all living forms on the
Earth.

As for that decision between 'left-handed' or 'right-handed' amino
acids, I dare say there had been many tossings of that coin, and many
different decisions. But, because all life now is descended from a single
common ancestor, it was only one tossing of one coin that was to
be remembered.

The logic of Darwin's tree makes the lateness of the last common
ancestor readily understandable. Given that new species can only
arise from pre-existing species through branching processes, and
given also that the vast majority of species become extinct, then it
becomes virtually inevitable that after a sufficient time all living

species will be traceable to some point well above the initial branch-point. Darwin's tree, you see, is something like this:

You can try drawing such a tree yourself – branching at random, pruning at random – and you will see that this is the usual sort of outcome: all the topmost twigs, corresponding to here-and-now organisms, are connected to some main branch-point well above the ground.

If there is an apparent absence of really primitive forms of life on the Earth, this is not because there never were such forms. It is not even mainly that primitive forms would find the competition too hard (although I dare say they might). No: a sufficient reason is already there in the logic of the sort of tree that Darwin described, where branching and pruning and branching and pruning...have been going on together.

Why are (some) 'molecules of life' easy to make?

There is a growing doubt about the idea that the primitive oceans would have been full of organic molecules. As discussed in chapter 1, it seems now that the early atmosphere of the Earth was dominated by nitrogen and carbon dioxide. This would have made the synthesis of organic molecules much more difficult than under the methane-dominated atmosphere that had previously been imagined.

It is being realised too that ultraviolet sunlight is even better at destroying middle-sized organic molecules than at making them. A general effect of ultraviolet light is to break covalent bonds. While this will tend to lead to the making of a wider spread of molecules – because the broken bonds will re-form in new ways – the general effect is nevertheless destructive. It is an atom-shuffling effect. The typical outcome is either a very complicated mixture (a tar) or simple, rather stable molecules like carbon dioxide and water (or, very often, first the one and then the other).

The remote-controlled landings on Mars by the Viking spacecraft served to emphasise the bleak side of ultraviolet sunlight from the point of view of chemical evolution. There are seemingly no organic molecules on the surface of Mars – because of the ultraviolet light. Indeed organic molecules are destroyed before they are made: the ultraviolet sunlight converts surface minerals into materials that will destroy organic molecules even more effectively than will oxygen gas.

Even without ultraviolet light or oxygen, mixtures of organic molecules do not keep very well. (Vintage port in its cold, dark cellar, changes over mere decades.) Complex mineral surfaces can have accidental catalytic effects that tend to accelerate the arbitrary shuffling of covalent bonds leading to more and more complex tars. If you want an example of how the Earth processes organic molecules, then look at raw petroleum. Organic minerals are usually exceedingly complex shuffled-up mixtures of this sort.

If indeed those active ('wound-up') little molecules cyanide and formaldehyde had been present in primitive oceans, they would have made matters worse. True, they are kicked into existence from more stable materials by various energy sources; and if you have them pure they will need no further pushing to make a few of the simpler 'molecules of life' for you (as well as many other things). But the waters on the primitive Earth would not have been pure. There would have been millions of kinds of organic molecules there if there had been any kinds at all. There would have been millions of ways for cyanide or formaldehyde to react. The result would have been a thicker, darker sort of tar.

Organic chemists are only too familiar with tars, gludges and gunks. Infernally complicated mixtures are only too normal products of organic chemical reactions. It is the sheer richness of ways of putting together carbon, hydrogen, oxygen and nitrogen atoms that creates the problem. There has to be much contrivance and control if any particular molecule of any great complexity is to be made in more than the minutest amounts. Even then, complicated mixtures are seldom avoided entirely. Most of the hard work in the synthesis of a particular organic compound is 'work-up'. This is a kind of weeding operation, after a reaction has taken place, in which molecules that you did not want are removed. To synthesise a molecule of any complexity usually requires many reactions in a row – with careful 'work-up' at each step, because, generally speaking, the product from one reaction should be pure before the next step is taken.

It is true that some of the simpler amino acids have been found in complex mixtures generated under conditions simulating those that might have been present on the primitive Earth. Even nucleotide letters have been found in mixtures that are said to be plausible simulations of probiotic products. But all such 'molecules of life' are always minority products and usually no more than trace products. Their detection often owes more to the skill of the experimenter than to any powerful tendency for the 'molecules of life' to form.

Sugars are particularly trying. While it is true that they form from formaldehyde solutions, these solutions have to be far more concentrated than would have been likely in primordial oceans. And the reaction is quite spoilt in practice by just about every possible sugar being made at the same time – and much else besides. Furthermore the conditions that form sugars also go on to destroy them. Sugars quickly make their own special kind of tar – caramel – and they make still more complicated mixtures if amino acids are around.

In sum the ease of synthesis of 'the molecules of life' has been greatly exaggerated. It only applies to a few of the simplest, and in no case is it at all easy to see how the molecules would have been sufficiently unencumbered by other irrelevant or interfering molecules to have allowed further organisation to higher-order structures of the kinds that would be needed: message tapes, selective control structures, etc.

Finally, even if I am wrong about all this and primitive geochemistry had shown a precision in organic reaction control quite unlike modern geochemistry; even if it had produced all 'the molecules of life' and nothing but 'the molecules of life' in ample amounts; even then it would still only have reached the edges of the real problem as outlined in the first four chapters. Still, somehow, an evolving machine had to be made.

But, you may say, there is something here that needs explaining. It is surely not a coincidence that sugars *do* form from formaldehyde; that adenine *does* form from cyanide; that amino acids *are* made preferentially in simulated thunderstorms – and that they turn up in meteorites. Can we not still say that there is something especially ubiquitous about at least some of 'the molecules of life'? They turn up all over the place.

Yes indeed; but let us get these things in perspective. Here are the brutal facts as they now appear to us:

1. Only *some* 'molecules of life' are 'ubiquitous'.
2. Most 'ubiquitous' molecules are *not* 'molecules of life'.
3. 'Molecules of life' are usually better made under conditions *other* than those characteristic of the early Earth.

The correct inference is that some classes of organic molecules are easier to make and/or more stable than others, and that our (highly evolved) biochemical machine incorporates a number of members of such classes. This is simply not surprising enough to be informative.

It is not surprising that the set of molecules that was eventually to be fixed in our biochemical system should have included some that were not too difficult to make and reasonably stable once made. That would have been good economical engineering, an understandable outcome in any case – whether such molecules had been prerequisites for, or products of, that early phase of evolution that established our central biochemistry.

What is much more significant, I think, is that nucleotides and lipids, which are crucial to our present system, are absent from the class of 'ubiquitous' molecules. Nucleotides and lipids have yet to be made under conditions that are realistic simulations of primitive Earth conditions. Nucleotides and lipids are much too complicated and particular for this to be surprising. They have all the appearance of molecules specially contrived for particular purposes. They have all the appearance of being, specifically, products of early evolution, not prerequisites.

Perhaps you still feel that 'time, and more time, and the resource of oceans' could have overcome the problems of how the more complex 'molecules of life' were originally made. I will now try to dispel such optimism by considering in more detail the most critical of all 'the molecules of life'.

Nucleotides are too expensive

The Sigma Company is one of several that compete to supply biochemicals for research purposes. Looking through their catalogue I find that I can buy a gram of ATP – a primed ('wound-up') RNA nucleotide – for about £5. ATP is only as cheap as this because it is relatively easy to extract from bulk biological materials – horse meat to be more specific. The other three primed RNA nucleotides are about ten times the price, and the primed DNA nucleotides cost about £300 per gram. But even these are only as cheap as they are because they are derived from natural biological materials.

As with postage stamps the price of nucleotides rises steeply with more abnormal types. The version of ATP with the sugar arabinose in the connector piece in place of ribose comes in at about £6000 a gram. But even such abnormal nucleotides, if they are synthetic (man-made) at all, are never wholly synthetic. Their manufacture will have started with components such as ribose obtained from biological sources. Usually the letter piece too will have been got that way. So £6000 a gram (or if you prefer £6M a kilogram) is a low estimate for the cost of a primed nucleotide 'in the open Universe' as it were. What would these materials cost if it were not for the horses (and others) that do most of the hard work? What would it actually cost to manufacture primed nucleotides from methane, ammonia and phosphate rock? I hate to think.

Contrast glycine and alanine, the two simplest amino acids. These really can be said to be easily made – they have been detected frequently in complex mixtures from sparking experiments, in meteorites, etc. Glycine comes in at about 1p a gram, and alanine (as a mixture of 'left-handed' and 'right-handed' forms) about 8p. (I may say that at these prices you get 99% pure material; thunderstorm simulations give you 99% *impure* material.)

Not only are they difficult to make, but primed nucleotides are rather unstable. Sigma recommend that the DNA primed nucleotides should be shipped in dry ice to avoid decomposition in transit, and nucleotides generally should be stored at below freezing point.

Expensive and fragile, primed nucleotides (or unprimed ones for that matter) are, I think, implausible as significant geochemical products – as minerals – at any time.

In *Genetic Takeover* I listed 14 major hurdles that would have to be overcome for primed nucleotides to have been made on the primitive Earth – from the build-up of sufficient and separate concentrations of formaldehyde and cyanide to the final 'winding-up' of the nucleotides. In practice each of these processes would be subdivided into separate unit operations that would have to be suitably sequenced.

In carrying out an organic synthesis in the laboratory there are tens or hundreds of little events: lifting, pouring, mixing, stirring, topping-up, decanting, adjusting, etc., etc. There may not be much to these unit operations in themselves, but their sequencing has to be right. There is a manufacturing procedure that has to be followed, and when such a procedure is at all prolonged it becomes absurd to imagine it being carried out by chance. That is why simple amino acids are plausible probiotic products, primed nucleotides are not.

It is not that one cannot imagine plausible unit processes on the primitive Earth that, taken together, might have yielded primed nucleotides – as one can imagine a coin falling heads a thousand times in a row.

Yes, you can imagine the primitive Earth doing the kinds of things that the practical organic chemist does. You can see a pool evaporating in the sun to concentrate a solution, or two solutions happening to mix because a stream overflows, or a catalytic mineral dust being blown in by the wind. You can imagine filtrations, decantations, heatings, acidifications: you can imagine many such operations taking place through little geological and meteorological accidents. But to show that each step in a sequence is plausible is not to show that the sequence itself is plausible.

But, you may say, with all the time in the world, and so much world, the right combinations of circumstances would happen some time? Is that not plausible?

The answer is no: there was not enough time, and there was not enough world. Let me try to justify this.

It would be a safe oversimplification, I think, to say that on average the 14 hurdles that I referred to in the making of primed nucleotides would each take 10 unit operations – that at least 140 little events would have to be appropriately sequenced. (If you doubt this, go and watch an organic chemist at work; look at all the things he actually does in bringing about what he would describe as 'one step' in an organic synthesis.) And it is surely on the optimistic side to suppose that, unguided, the appropriate thing happened at each point on one occasion in six. But if we take this as the kind of chance that we are talking about, then we can say that the odds against a successful unguided synthesis of a batch of primed nucleotide on the primitive Earth are similar to the odds against a six coming up every time with 140 throws of a dice. Is that sort of thing too much of a coincidence or not?

There are 6 possible outcomes from throwing a dice once; 6×6 from a double throw; $6 \times 6 \times 6$ from a triple throw; and 6 multiplied by itself 140 times from 140 throws. This is a huge number, represented approximately by a 1 followed by 109 zeros (i.e. $\sim 10^{109}$). This is the sort of number of trials that you would have to make to have a reasonable chance of hitting on the one outcome that represents success. Throwing one dice once a second for the period of the Earth's history would only let you get through about 10^{15} trials: so you would need about 10^{94} dice. That is far more than the number of electrons in the observed Universe (estimated at around 10^{80}).

Of course you might argue that in practice a synthesis might be carried through in different ways, and that is true, but remember what generous allowances we made in cutting down the actual amount of sheer skill that organic synthesis requires. And remember too that a manufacturing procedure is not usually very forgiving about arbitrary modifications: it all too easily goes off the rails never to recover. This is especially true of chemical processes, where it is usually not good enough to add the acid at the wrong time or throw away the wrong solution, or even use an ultraviolet lamp of the wrong sort. Careless organic synthesis only works when the product that is wanted belongs to that inevitably small set of molecules that are especially stable – molecules like carbon dioxide and water, even perhaps glycine and adenine in a much more limited way. But nucleotides are not like that to judge from the price.

One's intuition can lead one astray when thinking of the role of vast times and spaces in generating improbable structures. The moral is that *vast times and spaces do not make all that much difference to the level of competence that pure chance can simulate.* Even to get 14 sixes in a row (with one dice following the rules of our game) you should put aside some tens of thousands of years. But for 7 sixes a few weeks should do, and for 3 sixes a few minutes.

This is all an indication of the steepness of that cliff-face that we were thinking about: a three-step process may be easily attributable to chance while a similar thirty-step process is quite absurd.

Dicing with life

Intuitively one might have supposed that a thirty-step process would take about 10 times as long to be realised by chance as a three-step process. There are situations in which this would be the case: but only when there is a memory of success and failure in the past, where success can be built on. The dice-throwing analogy here is where we are allowed to continue throwing at each step until a six comes up and then go on to the next step – where we can accumulate the successes. That way 140 sixes could be reached with one dice in about 140×6 seconds, or about a quarter of an hour. But that kind of dice game only becomes possible in Nature where there are successions of organisms; where messages are being handed on; where a design is not spoilt by a single failure – where past success can still be built on in spite of failures because there are many copies of the past success in existence. It is then that biology begins. Dicing with organisms is a different game altogether.

But how did Nature start to play this game? At the very least a maintained supply of primed nucleotides would be required for any kind of organism using our kind of message tapes. A nucleotide-making factory would be needed. Surely only natural selection could have generated such a thing in the first place? Surely, then, there were earlier organisms that did not need nucleotides, but could evolve to produce them?

The way to surmount that cliff-face is to avoid it. There must be some other path to the mountain.

'The odds are enormous against its being coincidence. No figures could express them.'

A clue in a Chinese box

'That is another line of thought. There are two, and I beg you will not tangle them.'

A poodle, a Pekinese, a borzoi, a Yorkshire terrier – these all look very different from each other, yet we recognise that each is, in a sense, the same animal dressed up differently. To be members of the same species is to have a very similar biological organisation: animals that can interbreed must be of essentially the same design since the construction manuals in their Libraries must be, in that case, largely interchangeable.

Even between somewhat more distantly related animals, say between a whale and a bat, the design similarities may be far greater than the differences that catch the eye. Whales and bats are each of them mammals derived from a common ancestor not all that far back, and sharing most of the design ideas on which their survival depends. The machinery for breathing, digesting, excreting; the lay-out of nerves and circulation; the means of making skin and bone; the kinds of protein molecules – all these, and many more, are more similar than different between the whale and the bat. These two animals might also be said to be the same animal dressed up differently; even if this time you would have to go a bit deeper to see it. You would have to go deeper still to see what is the same between a herring and hamster, but the similarities are still not that deep. Still there are more similarities than differences.

Between a poodle and a petunia? The higher-order design features are certainly a bit different. There is nothing like a poodle's curly hair in a petunia; nor is there a liver, or bones, or the smallest tendency to persistent barking. Yet go deep enough and you will find, for example, a somewhat similar design of cells and some very similar protein molecules. Go deeper still and you will find those features that are common to all living things on this planet whether elephant, edelweiss or *E. coli*: you will find that most central biochemical machinery that we talked about in chapter 4, with always the same message tape material in the books in the central Libraries; with the messages being printed off in the same way, to be passed on to subsequent generations of cells; with the same main code used in its translation, and exactly the same set of amino acids that form the links in the protein chains that are made...

In the forms of life on Earth the variety is mainly superficial: as you go deeper there is less and less variety. Right at the centre there is none.

People often talk in this way about organisations of different sorts – in terms of levels. Some aspects are said to be superficial, while others are said to be deeper, still others central. This is a kind of Chinese box metaphor. The organisation is seen to have a kind of box-within-box structure. Outer subsystems rest on inner ones and everything rests on a central core.

Let us use this metaphor for organisms. Let us indeed think of the Earth as a museum housing an enormous and seemingly very varied collection of sets of Chinese boxes. How was this collection built up? How did it come about that the outer boxes of sets should be so much more varied in their style than the inner boxes? How is it that on taking apart any of the nested sets of boxes in the museum we find, always, the same central core?

It seems clear enough that the sets of boxes must have been built, by and large, by the addition of new boxes on the outside. This fits with our general understanding of sequences of events in evolution – more central design ideas are older. It is also common-sense. What we *mean* by a superficial design feature is that nothing much else within the organisation depends on it. It is much easier to add to the outside of an organisation, or to modify features that are on the outside. Darwin makes the point, in *Origin of Species*, that recently evolved features in organisms are indeed found to be more variable, in many cases, than features that have been present for a long time.

One can thus see, perhaps, how it is that the outer boxes of the sets in the museum are so variously painted. And one can see too why the inner boxes are much less easily modified *now*. But that still does not answer the question of how it is that there are fewer kinds of boxes deeper in, and only one kind of core. After all a box at any level was at one time an outer box. *Then* it should have been easily modified by evolution: then there should have been a great variety of boxes at that level. Why did this variety not persist?

Three standard ideas alongside each other can provide an explanation. Given the idea (i) that inner boxes become fixed as new boxes are built around them, we can add two ideas from the last chapter (pp. 41–2): (ii) that new species can only arise through branching processes from old (a new box set can only be a modified copy of a set already in the museum) and (iii) that most species go extinct (the Museum is always being plundered by thieves and vandals; whole wings get burnt down; cartloads of boxes have to be thrown away each year as a result of rain damage, woodworm, earthquakes, floods...).

Because of the stifling effect of the outer boxes on the variability of inner ones, and because of (ii), there is no way of making new central designs. Yet species keep on going extinct at a great rate. Central design ideas can only be lost, then, and the more central they are the longer they have been exposed to this hazard. Hence the outcome observed in the museum collection of organisms on the Earth – highly variable exteriors with increasingly conservative interiors.

Beyond the last common ancestor

Given all that, it is not at all surprising that the central common core should be so complex – if you also look again at the tree on p. 42. The common biochemical machinery to be traced back to the last common ancestor of life now on Earth is precisely that core in all the sets of Chinese boxes in the museum. As discussed in the last chapter, the enormous complexity of this invariant core demands some extended evolutionary process for its making, some process between the true base of the evolutionary tree and that branch-point, far from the base, that represents the last common ancestor of all life now.

Now look at the tree more carefully. You will see that the position of the last common ancestor in the tree is an accident of our point of view. The position would have been at a lower branch-point in relation to organisms that had lived long enough ago. (Cover the top half of the tree with a piece of paper and you will see that the new branch tops that you have made are now connected together at a point below the previous connecting point.)

During the early evolution of life there would have been a whole succession of last common ancestors. The question is this: Were the cores of earlier ones smaller than the cores of later ones? Was that how our immensely complex common core evolved – through processes analogous to later processes, with outer shells fixing inner shells? If that is what happened, then our common biochemical core should reveal its history in its structure. It should have a box-within-box structure; it should be *nested*. And the nature of its nesting should let us see a sequence of evolution beyond the last common ancestor of all life on Earth. That should be worth seeing!

Dilemma

Is the central biochemical system of all life on the Earth nested?

Yes, but, what is a little confusing, it is nested in two different ways. What is worse, the different ways suggest different sequences of events in early evolution. Let me explain.

According to one way of looking at it, the innermost box of the central core is DNA. This is the box that contains the quintessence of the organism, the ultimate controller, the genetic information. Then, outside the DNA, is the RNA box, and outside that the protein box and boxes corresponding to more or less direct products of the activities of proteins – more distant control structures such as membranes. Such layering is easily visible within the core. It can be seen by asking the question 'What is needed to control what?' The answer is always, in the end, DNA – suggesting that DNA came first.

But there is another way of looking at an organisation: not through the boardroom, through *the control structure*, but at the lay-out on the factory floor. Not by asking the question 'What is needed to control what?', but by asking instead 'What is needed to make what?' We can call this *the supply structure* of an organisation.

The supply structure of the central core of organisms is to be found, more or less, in what are called the primary metabolic pathways. These are the sequences of procedures used in assembling and disassembling such molecular micro-components as amino acids, nucleotides, lipids etc. – those now universal 'molecules of life'.

The organisation of these primary metabolic pathways is somewhat like the organisation of roads in a typical English market town. This supply structure of the central core is manifestly nested, to at least one level, in that the core has itself a centre, a kind of commercial centre or market-place, a region where essential goods can be bought and sold within easy walking distance. Then radiating from this commercial centre are the main roads (some one-way, out or in; others two-way).

This centre of centres, this biochemical market-place, deals in subcomponents. These are small molecular pieces into which the generally somewhat larger 'molecules of life' are disassembled and from which these larger molecules can also be made. In higher animals, such as ourselves, a number of the manufacturing routes have fallen into disrepair; but broadly speaking any of the 'molecules of life' can be made from any other through a suitable combination of takings apart and puttings together – by going into the central region along one route and going out again along another.

What are these most central go-between molecules? There are about a dozen of them: all contain carbon, hydrogen and oxygen atoms; a few also have a phosphate group in them. One, called acetyl (really only part of a molecule), is commonly held on to through a sulphur atom.

The atom curiously missing from these regions is nitrogen. This is a striking difference between the supply structure and the control structure. Nitrogen is ubiquitous at the control centre – in DNA, RNA and proteins.

In the supply structure, amino acids are at least one box out from the central region of the core. Eight of these can be made quite easily from central subcomponents (by adding nitrogen in the form of ammonia among other manipulations). They can be seen as constituting a distinct shell in the supply structure. Nucleotides are much further out, requiring among other things two of the inner amino acids for their manufacture. Nucleotides are really quite far away from the centre. About a dozen and a half separate operations, involving as many enzymes, are needed to make one of the DNA nucleotides. Yet all this is still within a core in the supply structure that is common in its essentials to all organisms now on the Earth.

Because of the easy interconvertibility of the most central subcomponents, it is difficult to locate an exact centre – as it may be difficult to decide about the exact centre of a town. (Should it be the Town Hall, or the Post Office, or the War Memorial, or the King's Arms?) It is likely to be a somewhat academic question in either case, the real centre being a region rather than a definite spot. Nevertheless if we want an equivalent to DNA for the supply structure of organisms – if we want a single substance to put right at the centre of the nested boxes – then I suppose it should be carbon dioxide. Not that this is the immediate source of carbon for all organisms, but it is the source for plants and hence the ultimate source for all organisms: and if it is not actually located at the centre even for plants (the principal supply point is slightly off the main street), all those central subcomponents are fairly closely related to carbon dioxide chemically. (They have a relatively high proportion of oxygen atoms in them.)

We are left with a dilemma. An examination of the control

structure of organisms suggests that DNA was the first substance for life, while an examination of the supply structure leads to a quite different conclusion – that in the beginning there was something rather simple that had no nitrogen in it, something like carbon dioxide.

Which to believe?

Here are sketches of the 'control core' and the 'supply core' for organisms now on the Earth:

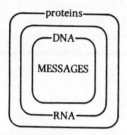 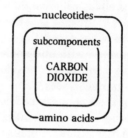

Which contains the true clue? Or does neither? Or do, somehow, both? Let us now consider two hypotheses.

According to the C-hypothesis (C for control) the true secret is in the control structure of the common core of organisms. DNA (or something like it) has always been at the centre, the control machinery evolving outwards with new boxes, new remoter means of control, being added on the outside.

To begin with, DNA-like molecules were selected by their surroundings directly according to the sequences that they happened to be holding. The selection was rough, and the exact sequences of letters did not matter very much. Nevertheless some sequences were better than others. For example certain sequences caused the molecules holding them to fold up into a compact ball that protected them against destruction in certain sorts of places. So in such places such sequences would be found more often, and they would be found to be catching on more and more.

There is certainly nothing logically wrong with the idea of evolving DNA-like molecules when you allow that a message in a particular such molecule can have an effect on the properties of that particular molecule. This has indeed been demonstrated in the laboratory for one DNA-like molecule – actually RNA. With the help of a suitable big enzyme, and a supply of wound-up nucleotides, RNA molecules can be made to replicate in the test-tube and, if the conditions are right, to evolve.

Now given evolving DNA-like molecules, you could imagine other ways in which messages within them could be effective – more indirect ways via effects on other molecules, such as amino acids, and eventually via the ability to join together amino acids to make proteins. Out and out, box without box, the control structure can be seen building up to make the now central core; and then, after that, out to further boxes to create all that immense variety of indirect means through which DNA now contrives its own propagation.

This is all logical enough. But is it true? Remember the appalling difficulties in the idea that the Earth ever manufactured nucleotides. And then what about the supply structure? Why does that tell a different story, with nucleotides coming in so late? What guided the evolution from carbon dioxide (say) towards those crucial, difficult, wound-up nucleotides?

Again there seems to be a logical answer – if you can believe that there were supplies of nucleotides on the early Earth. You can imagine that as these supplies began to run out, organisms that could make them from somewhat simpler things had an advantage: then, as these somewhat simpler components ran out, there was a race on to make them from still simpler, more available materials. And so on, all the way to the simplest and most available source of all – carbon dioxide. According to this story the supply structure of the central core was built in reverse, from the outside in.

Again, though, is this true? There are great difficulties over and above the whole idea of wound-up nucleotides having been there in the first place. It takes a dozen and a half steps to make a nucleotide – there are that many intermediates, many of which are quite unstable. It is not at all clear that these intermediates would have been available for use in a primordial soup even if primed nucleotides had been.

The C-hypothesis, for all its logic, is unsatisfactory when faced with practicalities.

Consider, then, the S-hypothesis (S for supply), that the real clue to the origin of life on Earth is to be found in the other set of Chinese boxes, in the supply structure of the common central core. Here, then, is another story.

Carbon dioxide has always been at the supply centre, with nucleotides and then the nucleic acids, DNA and RNA, coming in late. Evolution was always in the normal outward direction, from simple supplies to more complex products. The DNA control structure too was built outwards, but this whole phase only started quite late on after internal nucleotide supplies had been established.

There is little question that the most straightforward reading of the biochemical map puts carbon dioxide early and nucleotides late. The great biochemical explorer Fritz Lipmann pointed this out some twenty years ago. More recently Hyman Hartman developed the very hypothesis that we are now discussing (calling it 'the onion heuristic'). But, as Hartman saw, this S-hypothesis carries with it an essential rider. As there can be no evolution – of pathways or anything else – without replicating messages, without forms that can be copies of copies of copies..., there has to be *some* sort of a genetic material in *any* sort of organism. If it was not nucleic acid to begin with, then it must have been something else. We would have to say, then, that before the nucleic-acid-centred control machinery there was another kind of control machinery. We would have to say that there were earlier kinds of organisms that did not need nucleotides, but could evolve to produce them. And *that* conclusion we came to at the end of the last chapter.

> 'When you follow two separate chains of thought, Watson, you will find
> some point of intersection which should approximate to the truth.'

8

Missing pieces

'I don't say that we have fathomed it – far from it – but when we have traced the missing dumb-bell –'
  'The dumb-bell!'
  'Dear me, Watson, is it possible that you have not penetrated the fact that the case hangs upon the missing dumb-bell? Well, well, you need not be downcast; for between ourselves I don't think that either Inspector Mac or the excellent local practitioner has grasped the overwhelming importance of this incident. One dumb-bell, Watson! Consider an athlete with one dumb-bell! Picture to yourself the unilateral development, the imminent danger of a spinal curvature. Shocking, Watson, shocking!'

In one way the eye is eminently understandable. It is so like a camera that you wonder why there is not a law suit going on somewhere for breach of patent. The dark box, the lens, the iris diaphragm, the light-sensitive surface – each of these components is there in each case. At deeper levels there are certainly patentable differences in design. The light-sensitive area at the back of the eye is not actually much like a film. It, and many other things about the eye, are not by any means fully understood. But what is eminently understandable about the eye is that it should consist of rather definite components working in collaboration: as remarked in chapter 6 (point two, to which we are now returning) this is what really efficient pieces of machinery are usually like.

The bit that is not so clear about the eye – and a favourite challenge to Darwin – is how its components evolved when the whole machine will only work when all the components are there in place and working.

Not that this problem is peculiar to the eye. Organisms are full of such machinery, and it is a widely held view that this appearance of having been designed is *the* key feature of living things. (Recall Coleridge's definition of life as 'a whole that is pre-supposed by all its parts'.)

How can a complex collaboration between components evolve in small steps?

The first thing to notice is that a structure within an organism often – usually – has several different functions. An animal's jaws, for example, may have several uses other than for eating – for fighting, for carrying young, etc. It is clear that not all such functions were hit on at once. Some would have been later discoveries. If new uses may be found for old structures, so too can old needs be met by more recently evolved structures. There is plenty of scope for the accidental discovery of new ways of doing things that depend on two or more structures that are already there. For example, the cat's way of keeping warm by means of a furry coat is perhaps only a good one if there is a way of keeping the coat clean. No one would say that the tongue evolved originally for this purpose: but it turned out to be useful all the same as an essential part of a Fur Insulation System. The scratchy cat's tongue is now modified for cleaning purposes, as well as, still, carrying out its more ancient role as part of the Food Processing System. And, of course, the tongue has other uses too. It helps some animals to keep cool, others to speak, and so on. This is all very typical at all levels of organisation, from organs to molecules. There are components in organisms that have many uses that cannot all have been original uses; there are components that depend on each other in ways that cannot have been original, and yet may now be vital.

The fact is that even the so-called simple organisms such as *E. coli* are very complex enterprises with all sorts of things going on together. There is plenty of scope for accidental discoveries of effective new combinations of subsystems. It seems inevitable that every so often an older way of doing things will be displaced by a newer way that depends on a new set of subsystems. It is then that seemingly paradoxical collaborations may come about.

To see how, consider this very simplified model – an arch of stones:

This might seem to be a paradoxical structure if you had been told that it arose from a succession of small modifications, that it had been built one stone at a time.

How can you build any kind of arch *gradually*?

The answer is with a supporting scaffolding. In this case you might have used a scaffolding of stones. First you would build a wall, one stone at a time:

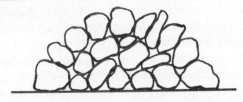

Then you would remove stones to leave the 'paradoxical' structure. Is there any other way than with scaffolding of *some* sort? Is there any other way to explain the kind of complex leaning together of subsystems that one finds in organisms, when each of several things depends on each other, than that there had been earlier pieces, now missing?

Nowhere is a collaboration of components tighter than in central biochemistry. Pull out a molecule – any molecule. What is it? Aspartic acid? That is as good an example as any. Aspartic acid is one of the twenty protein amino acids. It is, then, a component of virtually every enzyme. Every chemical reaction in the cell to this extent depends on aspartic acid being there, which means that every molecule made by the cell depends on this molecule. But, as is so often the case, this molecule is also used as a building block for all sorts of other molecules – for some of the nucleotide letters for example. And of course nucleotide letters are of central importance... Pull out another molecule, any one of the central set, and ask 'What use is this?' and you will find the same thing: you will find several immediate answers, and then, when considering more indirect effects, you will find that every molecule is required in some way or other by every other molecule.

It is a far more complex architecture than an arch, because one 'stone' does not connect only with two others, but with many: it is an arch in many dimensions and the more unchangeable for that. Nothing can be touched or the whole edifice will collapse. Looking at the structure of interdependences in central biochemistry it is not at all difficult to see why central biochemistry is now so fixed and has been for so long. The difficult question is how such a complexity of arching evolved stone by stone.

Think of the sheer amount of evolving that had to go on in making all that machinery needed to manufacture protein molecules (described towards the end of chapter 4). Think of all the selectings and rejectings that are implied by the enormous sophistication and complexity of the outcome. The evolution of the code is only part of the problem – although it is one of the sharpest parts. Think of what an agreed code pre-supposes. Try to imagine the situation that would allow the evolution of anything so complicated and so fixed, and so seemingly inevitably fixed – and so utterly indispensable now – as our central biochemical machinery. Whatever that situation was it must have been very different from now.

Surely there was 'scaffolding'. Before the multitudinous components of present biochemistry could come to lean together *they had to lean on something else.*

We have been here before. We keep coming to this idea that at some earlier phase of evolution, before life as we know it, there were other kinds of evolving systems, other organisms that, in effect, invented our system.

But can one imagine any way in which control could be handed over from an old system to a new system? Even so, would we not simply have the same problem as before (only moved back, and further out of sight)? The answer to the first of these questions is 'yes', and to the second 'no'. Let us start on the first question.

The long rope

The main burden of chapter 2 was that evolution can be said to consist of the elaboration of genetic information. Admittedly for this to be possible the genetic messages must have some effect, they have to be expressed as hardware – they have to give rise to phenotypes of organisms. But all this is strictly the next part of the argument: the messages come first because only the messages as such have a long-term survival; only they give the long-term continuity to lines of succession; it is only they that can be said to evolve.

The lines of succession are not monofilaments. What is passed on from generation to generation is a bunch of messages, that is to say a bunch of genes. What evolves are bunches of genes, and the bunches can change not only through modifications to members, but also through additions and subtractions.

A long line of organisms, then, is like a rope made up, as most ropes are, of long overlapping fibres. It is not necessary that any fibre extends from one end of the rope to the other. Genes come and genes go.

I dare say there are telling administrative advantages in having all the genes in an organism made of the same stuff and operating the same way. Once you have a material like protein to work with – you can make almost anything out of protein – then such a sophisticated simplicity makes sense. We have only one life form to look at and, it is true, its organisms are always homogeneous (pun intended). But this is surely an incidental rather than a crucial point. We can see why it is a good idea; but we can also see that it could have been otherwise. The reasons why a bunch of genes may be successful do not require as a matter of principle that the genetic material should be uniform or that the modes of action of the genes should all be the same. What is needed is that together they can produce a phenotype that is to their mutual advantage. That is all that is needed.

There is then a simple way in which the central control machinery of organisms could have been updated: through a gradual takeover. A rope of hemp fibres at one end could gradually transform into a rope with only sisal fibres in it, by hemp fibres fading out and sisal fibres fading in. Similarly a line of organisms with genes made of one material might change gradually into a line of organisms based on a quite different genetic material, i.e. through a **genetic takeover**.

The idea of genetic takeover in early evolution does not simply displace the problem of the origin of life, because of course the earliest genetic control systems would not – could not – have had that paradoxical 'arched' organisation that has been causing all the trouble. It would have been 'scaffolding', buildable piece by piece. The problem of the origin of life becomes, above all, a search for those missing pieces – starting with the very first ones. What have we to go on?

Building up a picture

You do not have to have met someone to be able to build up a picture of what they are like – especially if you know where they come from and what they have achieved. Our ultimate ancestor is a bit like that. We have at least some strong suspicions about that first organism at the very base of the tree of evolution.

We are assuming (from chapter 1) that:

(1) *our ultimate ancestor was a product of the Earth*: it was made of components that the Earth provided. And we know by definition (although also by its achievements) that

(2) *it could evolve under natural selection*. This should be a good clue, since such systems must have very distinct properties, conforming to the abstract description of any organism given in chapter 2. From this present chapter we can add two other points, that

(3) *our ultimate ancestor's subsystems were not strongly interdependent* (i.e. not 'arched'), and that

(4) *genetic takeover provided a means of transition* to the decidedly 'arched' system that we know.

Genetic takeover also provides the possibility that our ultimate ancestors were made out of different materials from organisms now. This is as well, since we had decided (in chapter 6) that at least one of the kinds of crucial modern micro-components – the nucleotides – could not have been there in the very first organisms. From the last chapter too, it seems that nucleotides would not have been there for a long time. But how different chemically were those very first organisms? Even if they could have been very different are there any reasons to think that they actually would have been?

There is at least a strong suspicion that first biochemical materials were quite different from now. It comes from a consideration of a distinction between 'low-tech' and 'high-tech' design approaches. Let me clarify my use of these terms.

Sticks and stones would be archetypal 'low-tech' devices – for use as weapons, props, means of making fire, and so on. Even a spear is still 'low-tech' if it is simply a modified stick; or a tinder box if it is simply an elaboration of the idea that you can make sparks by banging two stones together. The question is (more or less): How soon, as it is being put together, can the machine start working? If there are very simple forms of a machine that will work then it is 'low-tech' in its basic design approach. Clearly, the very first organisms must have been 'low-tech'.

The 'high-tech' approach is quite different. Here the whole idea is that an overall function (say personal transport) is achieved through a collaboration of diverse components (things like pistons, rubber tyres, spark plugs, a tank of highly inflammable liquid...). Now whether any such 'high-tech' components can even begin to perform the overall function by themselves, or in any simple combination – or indeed whether any cut-down version of the machine will work at all – is irrelevant. The only criterion is efficiency. The only question is this: Once fully assembled does the machine work?

It is our experience that 'high-tech' machines are on the whole more efficient than 'low-tech' machines, even highly elaborate 'low-tech' machines. This is not really surprising when you consider what constraints must be imposed on a design by the need to be like some primitive machine. Freed from that functionally irrelevant constraint you can take it that designers of efficient machines will almost invariably choose 'high-tech': they will choose designs that by and large do not work until fully assembled. As indicated in this chapter, evolution too can achieve 'high-tech' designs: there are routes to such designs (remember the arch and the rope).

It is also our experience that 'high-tech' machines are usually made with different materials from their 'low-tech' counterparts. Again this is not surprising. It arises from the difference of approach. The much greater subdivision of the overall function of the machine into subsidiary functions also *changes* the functions: none of the components is required to do what the one or two components in a primitive 'low-tech' machine have to do. The parts now pre-suppose a complex whole, they are made to collaborate – and they are made differently because of that. The calculator is not made from the same materials as the abacus, nor is the machine-gun made from the same materials as the bow and arrow. (Readers's home project: think of sixty-seven other examples.)

The pieces of those first 'low-tech' organisms would have had different design constraints on them from the design constraints on any of the tightly interlocking components of the modern 'high-tech' machine. Hence the strong suspicion that

(5) *our ultimate ancestors were made from quite different materials from modern organisms,* although (from the last chapter)

(6) *evolved forms of 'low-tech' life were able to manufacture our present biochemical components* and also

(7) *carbon dioxide was the original carbon feedstock.*

Of these Seven Suspicions about our ultimate ancestor, the fourth and fifth are certainly the most libellous. But they have the character of a double key. Used together they open new possibilities.

> '...perhaps the scent is not so cold but that two old hounds like Watson and myself may get a sniff of it.'

9

The trouble with molecules

'It is of the highest importance in the art of detection to be able to recognize, out of a number of facts, which are incidental and which vital. Otherwise your energy and attention must be dissipated instead of being concentrated. Now, in this case there was not the slightest doubt in my mind from the first that the key of the whole matter must be looked for in...'

But *how* do you decide, Mr. Holmes, which facts are incidental and which vital? What are the criteria?

There are no set rules, and the pre-conscious mechanisms that we discussed in chapter 5, and which doubtless Holmes made good use of, can lead one astray. For our investigation, however, there is a rule of thumb. It comes from the proposition that biology is the study of the causes and effects of evolution. Well, in thinking about the origin of evolution, causes are more likely to be vital, and effects incidental. Hence our interest (in chapter 2 mainly) in the abstract nature of organisms, the abstract features that allow them to be subject to evolution through natural selection. Most vital is the ability to transmit potentially complex information to offspring – and to do this pretty accurately.

Above all what turned out to be incidental was the unity of biochemistry. This was a Big Red Herring. The unity of biochemistry was seen to be an effect of a quite substantial period of evolution.

Up to now this book has been very largely a sifting of the vital from the incidental. We have now a concentrated residue of supposition:

Evolution started with 'low-tech' organisms that did not have to be, and probably were not made from 'the molecules of life'.

The first part of this statement might seem rather obvious were it not for the baleful conclusion, from chapter 2, that the design of any conceivable organism is inevitably very very complicated – with robot machines that can make other machines (including ones like themselves) under instructions held in an information store that can be replicated by means of yet more machinery whose construction is also specified in the information store and can be executed by the robot machines...

But that was another Big Red Herring. It arose from the unstated assumption that you actually need any machinery at all in an organism. Once you think you will need any, then you will think that you need a lot. If, for example, the organism has to have some kind of printing machinery in it, so that it can replicate its genetic information, then it will need manufacturing machinery also to make this printing machinery. And then this manufacturing machinery, some sort of robot, must also be able to make other machines exactly like itself. The circle closes eventually, but not until after a long journey – too long to be a practicable piece of engineering even for us, and much too long for Nature before its engineer, natural selection, had come on the scene.

So why start on such a journey? Only the messages are in principle essential for evolution, although in practice there has to be a material to hold the messages and physical means for their replication. But the components for making the genetic material can be provided by the environment and so can any machinery that is needed to work with these components to bring about the replication of the messages. An organism need be no more than a **naked gene** if the environment is kind enough.

This point was indeed made, in chapter 7, when we noted that RNA molecules can evolve. The key idea is that although the success of genes nowadays may depend on at least some of the information that they carry being translated into action via elaborate machinery, such an indirect mode of action is efficient rather than essential in principle. It is likely, then, to be a later effect of evolution rather than a necessary pre-condition. In principle – and indeed in practice for RNA – a message in a gene may have a direct consequence on the properties of the gene itself. For example, the sizes and shapes of gene particles might be affected by their internal messages, these properties affecting the survival chances, or the ease of replication or spreading of the particles that have them. Harking back to chapter 1, one can be more exhaustive and general. As soon as you have structures of any sort that can hold and replicate specific patterns; and when such patterns are occasionally subject to arbitrary modifications; and when such modifications are also then replicated; and when the patterns in question can be of an immense number of different kinds, and where *what* kind can make a difference to the survival chances and/or replicability and/or propagation of the replicating particles – why then (deep breath) evolution through natural selection could hardly be prevented. There are a lot of 'ands' (which is good news really because they are going to limit the possibilities), but note: there are no robots among them.

But does this not simply shift the difficulty from the organism to the environment? Certainly it shifts the difficulty, but it does not *simply* shift the difficulty. The difficulty changes, and it becomes much less severe. There do not have to be robots *anywhere*. The environment might possibly have to provide some sort of printing or replicating machinery, but it would not have to provide another instructable machine to make such machinery. Indeed it is a matter to be decided whether the environment would even have to provide anything that could be called replicating machinery, or machinery of any sort.

There would be but three things that an environment would have to provide for 'naked genes':

(i) *material units* out of which new genes could be made (by template replication);

(ii) *conditions* that would allow this to happen (whether or not these conditions included any sort of replication machinery); and

(iii) *reasons* why some genes should do better than others (what are called selection pressures).

It is true that now for RNA, the material units are probably too complex as primitive Earth products; and it looks as if a big enzyme has indeed to be included under (ii). But these are incidental features, not vital. They are specific objections to RNA. They depend on particular attributes of RNA molecules – and, anyway, we had decided in the last chapter that neither RNA nor DNA was the original genetic material.

For those first naked genes we must look for something more down to earth to satisfy conditions (i), (ii) and (iii) – and this search will occupy us very considerably during the next four chapters. For the rest of this chapter we will be concerned with a still more fundamental and general question: How do atoms get organised? Genes, like other control structures, are necessarily made up of very many atoms put together in a definite way. How are structures of this sort of size assembled within organisms, and how are such structures put together most easily without organisms?

Molecular construction systems

It is both a virtue and a snag of organic molecules that there are so many kinds of them. There is quite a variety of ways of putting together even just ten or so atoms; but by the time we are thinking about organic molecules of around a hundred atoms, the number of possibilities has leapt towards the astronomical – and with a thousand atoms, away beyond that. Most of the key control structures in organisms are bigger still. A smallish enzyme, for example, contains some 5000 atoms put together in one particular way, while the piece of RNA message tape that specifies that enzyme has some 30000 atoms in it. A membrane, say a membrane surrounding a cell, is not quite such a tightly specified object: but it still has to be well engineered. Its making may require that a few billion atoms are appropriately organised, in this case through many protein and lipid molecules in association.

Organisms today can engineer such components. They can make use of the virtue in the fabulous potential variety of organic molecules to make devices that are just right for their purposes. They can do this because they have the competence to overcome the great snag about such a variety of choices – that there are so many, many ways of going wrong.

How do they do it? One can discern three tightly interwoven techniques. They do it partly by (1) *molecule-manipulation*, with specially designed machine-tools such as enzymes or ribosomes; partly by (2) *pre-arrangement*, by setting things up so that there is only a limited number of possible outcomes; and partly by letting the molecules put themselves together into higher-order structures – what is called (3) **self-assembly**.

The first of these techniques is, perhaps, the one that we can understand most easily. It is the nearest to our normal human techniques of manufacture as applied to large-scale objects such as socks or motor cars. A ribosome is obviously a manufacturing machine. It is in this respect highly understandable.

We may not notice it so much, but the second of these techniques also applies to human manufacturing procedures. In knitting a pair of socks the right ball of wool should be to hand; for robot-building a motor car it is imperative that the correct components are there for the skilful but stupid robots to work on. Enzymes or ribosomes are enormously skilful but quite stupid. They can only work in rigged surroundings, where, among other things, there is only a certain limited set of molecules available. Enzymes are easily confused by molecules that are close to, but not quite the same as the molecules that they are supposed to work on – and there are many other kinds of things that will inhibit or destroy enzymes. So there has to be pre-arrangement.

The third construction technique of organisms – self-assembly – is the least like human techniques of manufacture. It is the special province of objects that are small enough for their heat agitation to be a dominant factor. It is a very common way for multi-atom objects to get put together on the surface of the Earth. So let us look into this a bit further.

You may think, when you blow a soap bubble, that it is all your own work. But really it is the soap molecules that are mainly responsible. They 'self-assemble', billions and billions of them, to make that amazing object.

A soap molecule is somewhat like a long tadpole in its shape, having a tail consisting of a chain of carbon and hydrogen atoms, and a head that has two oxygen atoms in it and a negative electric charge. The tails stick fairly well to each other through weak secondary (non-covalent) forces, but they do not stick at all well to water molecules. Molecules or parts of molecules that are made up of carbon and hydrogen atoms like this are said to be hydrophobic, a quaint description which means literally that they hate water. The heads on the other hand just love water, and they are said to be hydrophilic. The electric charge helps here: it helps to form quite strong secondary bonds between the head groups of soap molecules and water molecules.

Now when you dissolve some soap in water, the molecules form little clusters in which the tails are wriggled up against each other on the inside (away from the hated water) while the heads are on the outside (in contact with the lovely stuff). These clusters of soap molecules are self-assembled objects. (And very useful they are too, because the molecules of greasy materials are also born water-haters, but love those clusters of tails – which is why soap-and-water dissolves grease. It's all done with self-assembly. They don't know anything about chemistry, these molecules, but they know what they like.)

The soap film is simply another way of solving a molecule relations problem. The tails are also quite content to stick straight into the air, especially if they are alongside others so that all the tails can be packed closely together sideways. Soapy water always has a covering of soap molecules like this, and a soap film is two such coverings back-to-back with a thin sheet of water between them. Another kind of self-assembly.

Organisms make good use of these love–hate relationships and it is particularly easily seen in the membranes that are around cells and within cells. These membranes are largely composed of soap-like molecules (actually two-tailed molecules with rather more elaborate heads) called membrane-forming lipids. Because of their well-designed shape they do not just form clumpy clusters in water but tend to line up alongside each other into well-regimented sheets – membranes. These are somewhat like soap films except that they have water on the outside and not air, and the tails of the molecules point inwards rather than outwards.

By far the commonest kind of self-assembled object in Nature is the **crystal**. The molecules in that sugar crystal, or that piece of ice, or that quartz sand grain, are assembled with incredible neatness. Yet there was no construction machinery. The molecules found how to pack together all on their own. It is commonplace and astonishing.

How is it that molecules assemble themselves?

Self-assembly works because molecules are in a constant state of heat agitation. They never stay still for a moment: but they also tend to stick to each other, and they stick better if aligned in certain ways (e.g. with hydrophobic groups together, with positive and negative electric charges close to each other, and so on). The perpetual motion of the molecules allows them to try out many possible arrangements and they can thus arrive at some particular arrangement that has the greatest possible cohesion. Such an arrangement, once found, will of course tend to persist.

That is by no means all that has to be said about self-assembly, although it gives the main idea. Here are some of the other conditions.

First, the temperature must not be too high – that is to say the heat agitation must not be too frantic – or the molecules will never be able to settle on anything.

Second, the temperature must not be too low or the molecules will take too long to find the most stable arrangement.

Third, the concentration of molecules must be high enough. There have to be enough molecules within sticking distance so that the rate at which they can find each other (so as to self-assemble) is fast enough to offset the rate at which already assembled structures are shaking themselves apart. (There will always be a certain amount of this going on at any temperature.)

Fourth, the molecules must be quite good at sticking together, that is the forces between them must be reasonably strong overall.

Fifth, the forces between the molecules must nevertheless be **reversible** – they must be of the tentative exploratory sort that we discussed in chapter 4. The molecules must not lock together on

contact. They must be able to keep on trying out different arrangements. Certain kinds of (strong) covalent bonds are suitably reversible,
as we shall come to discuss later: but within the domain of organic
chemistry (the chemistry of carbon compounds) the suitably reversible forces are almost invariably secondary forces, the forces
between molecules that are a tenth to a hundredth of the strength
of those covalent bonds that hold the atoms together within
molecules.

Sixth, there must not be too many sorts of molecule around.
Otherwise there may be so many possible ways of coming together
that there is no very distinct best way: that is to say no very distinct
form of assembly that gives in to the cohesive forces most effectively –
or if you prefer that best satisfies the love–hate relationships between
the component molecules. Even if there is some distinctly best
arrangement, if there are too many possible arrangements the rate
at which the molecules hit on the best one may still not be fast enough
to offset the perpetual disintegrating effect of the heat agitation.

All this adds up to 'self-assembly' being not exactly *self*-assembly,
but rather a kind of assisted or pre-arranged assembly. If you have
set up the conditions right, in particular if you have made a good
choice of molecules, then these molecules will 'self-assemble'.

This is as it should be of course. It would never do if molecules
just did what they liked under no sort of control. Higher-order
structures – often multimolecular structures – can be contrived precisely to the extent that their 'self-assembly' can be pre-arranged. It
is a wise and indolent kind of control, like *laissez-faire* economics: let
them do what they like – because things are so rigged that what they
will like is what is needed.

The most exquisitely pre-arranged self-assembly is to be found in
the folding up of a protein molecule. The pre-arrangement is mainly
in the order in which the amino acid links in the protein chain have
been joined together. If you remember (from chapter 4), this order
is a direct translation of a genetic message. It becomes, then, the most
extraordinary kind of message that you ever saw: it turns itself into
the thing that it is describing. According to the order of the rather
variously shaped amino acid units – which include water-lovers and
water-haters – the chain folds up in a particular way. This is the
particular way that best satisfies the cohesive forces between different
bits of the chain while putting the water-lovers (as far as possible) on
the outside and the water-haters on the inside. Thus a straggly chain
becomes a compact working machine – an enzyme for example.

When you blew that soap bubble you had engineered rather general conditions and then let the molecules do the rest. The construction system for making protein machinery depends similarly on a division of labour between pre-arrangement and self-assembly. If enzymes are more difficult to make than soap bubbles, that is because, among other things, the proportion of pre-arrangement in such a protein molecule is very much higher. Among those other things is that much of the pre-arrangement in the protein construction system is through the specification of a particular sequence of amino acids, done largely by molecule manipulation (I told you that the three construction techniques were tightly interwoven) and molecule manipulation is an especially tricky technique. It is not something that we as organic chemists are much good at. We have to be content with other grosser kinds of pre-arrangement most of the time: we bring about desired reactions by arranging large-scale conditions, dealing with billions of billions of molecules at a time, while true molecule manipulation is a one-at-a-time process of the sort carried out by an enzyme or a ribosome. And just look at the size and complexity of enzymes and ribosomes! They are not just big for the fun of it. They have to be big to do their jobs, to be able to select particular molecules from their surroundings and then act on them in particular ways.

The moral of all this is that if you are wanting a 'low-tech' construction system for multi-atom objects, self-assembly should be the major element with pre-arrangement next – and if you can do without molecule manipulation altogether, so much the better.

But is this a realistic proposal? Remember the great snag about organic chemistry that we discussed earlier: that in making organic molecules there are so many, many ways of going wrong. Doing organic chemistry is a bit like playing the violin – superb when well executed, but especially appalling otherwise.

It seems a pretty straight choice. Either you have a construction system which is flexible, allowing a large number of alternative things to be made (a 'violin' system); or you have a more limited 'tin whistle' system, more suited to the beginner, with fewer things to go wrong. It is not at all clear that organic molecules are right for 'low-tech', beginner systems.

A particular trouble with organic molecules is that they only self-assemble properly when they are fairly large. Only then will there be a sufficient overall cohesion between the molecules, or between the parts of a foldable molecule. (A soap molecule needs to have a long tail; a protein chain has to have some twenty units in it before it will start to fold up coherently.) But large molecules are difficult to come by, especially at the kinds of concentration and purity needed for precise self-assembly processes. The massive objections that there are to the idea that good supplies of nucleotides could have been pre-arranged by the primitive Earth (chapter 6) apply with a similar force to almost any organic molecule of that sort of size – the sort of minimum size needed for organic molecules to be able to self-assemble in water into higher-order structures...

It goes round and round. Is there some way of breaking out of this circle of complaints? What is incidental, and what is vital?

Get back to fundamentals. What was vital for the very first organisms was that they should have been able to evolve. There had to be messages, then, that could be passed between generations as copies of copies of copies... There had to be genes of some sort. Now any sort of gene must surely be a precisely made many-atom structure. Since the first genes could have had no pre-evolved machinery to help in their (template) replication, then the assembly of new units in this process must have been some sort of self-assembly. If this means that organic molecules were not involved, then so be it. We have come across plenty of indications already that the stuff of first life was different from that of life now. The substance as such was incidental; it was the kind of process that it took part in that was vital.

'...it seemed to me... that some new possibility had dawned suddenly upon him.'

10

Crystals

'It is an old maxim of mine that when you have excluded the impossible, whatever remains, however improbable, must be the truth.'

If organic molecules as a class fail to fit the specifications for 'low-tech' genes, what classes of material remain?

Start with this thought: the units for self-assembling first genes should have been *small*. The trouble with structures built from carbon atoms, if you remember, is that too much precise molecule-building is required before the point is reached at which precise self-assembly can begin to operate. We want materials that can get on with the self-assembly much sooner: materials built from small units which are easily produced under geochemical conditions and which, because they are simple, come in only a few types. This should alleviate those problems of pre-purification that are generally associated with self-assembly processes – all the more important because there would have to be many precisely assembled units in a gene that was to hold more than a trivial amount of information. And of course the units must be held together by reversible forces – strong kinds of forces, though, if the units are small.

Many small units held together in a precise way by strong reversible forces? That sounds like a description of an inorganic crystalline material. We will now close in on the idea that the very first genes were **crystal genes** – indeed that the very first organisms were inorganic-crystalline in nature, not organic-molecular as organisms are now.

Perhaps when you think of a crystal you think of a diamond, or a snowflake, or a piece of quartz. Pretty enough things these, but hardly very lively. Indeed so. You might be similarly unimpressed with the idea that the key structures in living organisms were organic polymers, if your view of organic polymers was only of such things as motor car tyres or plastic buckets. The crystal genes that we are looking for would be by no means just any crystals, as organic genes are not just any organic polymers. A crystal gene must be able to hold

74

substantial amounts of information, and replicate that information rather accurately through processes of crystal growth and crystal break-up. And the information that the crystal genes hold must have some effect or effects that help the genes holding that information to survive better, or replicate faster, or be spread around more widely.

Nevertheless quite ordinary crystals may show some of the essential attributes that we would be looking for. Here for example is an experiment that you can do that shows crystals at their liveliest – growing before your very eyes, putting themselves together, breeding even.

Take about 250 grams of photographer's 'hypo' (sodium thio-sulphate pentahydrate) and put it in a clean 250 millitre beaker together with 75 millitres of distilled water. Heat to near boiling, stirring the solution with a glass rod to be sure that it is properly mixed. Remove the beaker from the heat and cover it immediately with a loose-fitting lid to keep out the dust (a piece of glass is ideal) and leave strictly alone for several hours, for the solution to cool. With any luck nothing will happen. You had a hot solution, now you have a cold solution which may look no different. But really it is a magic solution now, all set to perform for you. Carefully take the lid off the beaker, drop one tiny piece of 'hypo' crystal onto the surface of the solution, and watch amazed at what happens. Your crystal grows visibly: it breaks up from time to time and the pieces also grow...Soon your beaker is crowded with crystals, some several centimetres long. Then after a few minutes it all stops. The magic solution has lost its power – although if you want another perform-ance just re-heat and re-cool the beaker.

This experiment illustrates two conditions required for the growth of crystals from a solution: **supersaturation** and **seeding**. Let us consider these one at a time.

Undersaturation – saturation – supersaturation

Suppose you start adding salt to a beaker of water. The salt dissolves to begin with, and so long as it continues to do so we say that the solution is undersaturated with salt. But there comes a point at which no more salt dissolves however much you stir it, however long you wait. The water is then said to be saturated with salt, forming a saturated solution. It is well known that some materials are more soluble in (for example) water than others. This is to say that saturated solutions of various materials in a given solvent (such as water) at a given temperature contain different amounts of those materials. 'Hypo', for example, is very soluble in water – you can get a kilogram to dissolve in a litre of water at room temperature; and so is common salt – about 370 grams will dissolve in a litre at room temperature. What about powdered glass? Insoluble, many people would say, but certainly not. It dissolves only slowly but a litre of a saturated solution of a silica glass has about one-tenth of a gram of silica dissolved in it at around room temperature. Even pure quartz sand, a crystalline form of silica, dissolves to the extent of about 6 parts per million, that is about 6 milligrams (thousandths of a gram) in a litre of water at room temperature.

Now to be supersaturated means to have more dissolved than there ought to be. There are various ways in which this situation can be brought about, but one of the easiest is to do what we did in the experiment: to make a solution at a high temperature, where the solubility is higher, and then reduce the temperature. 'Hypo' solutions are particularly easily caught out this way: the cold supersaturated solution almost literally did not know what to do. It had to be 'told' by adding a piece of crystal that already had its units (billions and billions of them) packed together in the way that is characteristic for 'hypo' crystals. The solution had to be seeded.

Clearly it must somehow be more difficult to start up a new crystal than to add to one that is already there. The reason is that very small crystals do not hang together very well – they are a good bit more soluble than bigger crystals. A solution that is supersaturated for big crystals may nevertheless be undersaturated for the very tiny crystals that would have to come first. More of this in a moment.

When we talk of a **crystal structure**, for example the structure of a 'hypo' crystal, we mean a particular arrangement of units in three dimensions, like goods packed in huge neat stacks in a warehouse, economically filling space, all stacked the same way. For 'hypo' the units are of three kinds: sodium ions, which are sodium atoms each with one positive electric charge; thiosulphate ions, which are molecules made from two sulphur atoms and three oxygen atoms and carrying two negative charges; and water molecules. These three kinds of units are in the proportion $1:2:5$.

It is amazing how quickly crystals can put themselves together (when you consider that the number of units in some of those 'hypo' crystals that you made exceeds the number of milliseconds in the whole history of the Earth). And the precision is amazing too, a precision of internal arrangement that shows itself in the familiar regular outline of a typical crystal. How can such precision arise from the haphazard collisions of molecules?

The key to the precision of this, as of all forms of self-assembly, is the reversibility of the forces that hold the units together. Even in a growing crystal the processes of growth and dissolution are both going on at the same time. Units are adding to the crystal from the solution; but at the same time units in the crystal are shaking themselves free under the effect of their heat agitation and through their attraction for water molecules in the solution. It is a dynamic balance. If the solution is saturated, that simply means that the rate at which units are being added to the crystals is exactly balanced by the rate at which units are coming away again. If the solution is supersaturated the addition process will be somewhat faster; if undersaturated, the subtraction process will win. Now of course the units have no eyes. They will often add the wrong way. When this happens the resulting bit of crystal will be destabilised; it will not hold together so well. So in that region the dissolving process will go a bit faster. Provided the level of supersaturation is not too high the chances are that this local dissolving of the badly made bit of crystal will be faster than the general rate at which new crystal is being formed.

The level of supersaturation is clearly important if this kind of error correction mechanism is to work properly. If the level is very high then the rate at which new crystal is forming by the addition of units from the overcrowded solution will be much faster than the average dissolution rate, and it may very well be faster than the rate at which some imperfect piece of crystal can re-dissolve. Another way of putting this is to say that while a solution must be supersaturated with respect to perfect crystal, it must nevertheless be undersaturated with respect to imperfect crystal if the error correction mechanism is to work. There is a pay-off: low levels of supersaturation will give only very slow growth rates, while higher levels will increase the rate of crystal growth but it will also increase both the number and the number of kinds of imperfections that are introduced into the growing crystals.

High enough levels of supersaturation will also give rise to 'spontaneous seeding'. There comes a point at which even very small and imperfect congregations of units can grow. Such congregations can arise by chance – most likely on surfaces of the containing vessel or on dust particles – to set off the processes of crystal growth without any added seed crystals.

Even with only a vague idea of how a crystal gene might hold and replicate complex information we can already see one condition that is going to be required: an appropriate, (probably rather low) level of supersaturation. This is because the assembly of new units into new crystal should be accurate, and it should only take place on well-formed crystal that is already there.

Another necessary (but insufficient) condition is that crystal genes should 'breed' in the simple sense in which crystallographers use that term: that is to say the crystals should break up as they grow so as to generate new seeds – as vividly demonstrated in the experiment with 'hypo'. The details of how this could happen in such a way as to preserve and propagate information will provide us with a number of leads to be followed up in chapter 12. In the meantime there are some more clues in this question of appropriate levels of supersaturation.

Continuous crystallisation

How do you maintain a given low level of supersaturation (for, say, a million years)?

Not in the way that we did the 'hypo' crystallisation. Here the supersaturation was high to begin with and fell away as the crystallisation proceeded. Then the process died on us. How could a crystallisation be kept going indefinitely?

We could have kept 'hypo' crystals growing for as long as we wanted if instead of a simple beaker we had set up what is called an open system – a vessel with inflows and outflows. More particularly we would want a vessel into which a solution, supersaturated to a given level, was continuously flowing and from which crystals in suspension were continually being removed. Such a thing is called a continuous crystalliser and is used for commercial production of crystals under constant conditions of crystal growth.

If our object is not actually to produce crystals but rather to keep a crystallisation process going indefinitely – if we are gardeners rather than farmers – then the sensible thing would be to put the outflowing material back into the system. The outflow suspension of crystals would be heated and re-cooled to provide the input super-saturated solution. (The whole set-up would now be a closed cycle, requiring only an input of energy to drive it, but any part of this cycle would be an open system.)

Now we would have not only growth and reproduction ('breeding'), but also mortality. Any crystal will sooner or later be carried out the waste pipe, the whole thing only keeping going through new crystals being formed at the same time – and only by reproduction if the supersaturation level is low enough.

Such a cyclic continuous crystalliser would be a microcosmic analogy for the whole of life on the Earth. Any organism you think of lives in an open system in a situation similar to the crystals in our cyclic continuous crystalliser. There has to be in each case a maintained supply of materials and energy – food or sunlight etc. for organisms, or a 'wound-up' (supersaturated) solution for the crystals.

One begins to see, perhaps, the general kind of situation required for simple organisms that depend on crystal growth processes. If they are to last very long they have to live in a continuous crystalliser. Many sea creatures today, for example, depend on seawater being supersaturated with respect to calcium carbonate (chalk). These creatures then ingeniously 'seed' the formation of calcium carbonate crystals in making their shells etc.

Such are quite recent organisms whose inorganic components are very much at the outer edges of their organisations – in their outer Chinese boxes. More sophisticated organisms may use energy to contrive suitable levels of supersaturation within themselves (as we do, presumably, for making our bones and teeth). But we can take it, I think, that by far the most favourable place for a first organism dependent on crystal growth processes would be in a continuous crystalliser – some solution that was being maintained (just?) supersaturated by appropriate inflows and outflows. The sea is like that, but so are many other places on the Earth, as we shall see.

If there is anything in the idea that a crystal of simple units could be a 'low-tech' alternative to DNA; if there is, or ever could be, such a thing as a crystal gene, then the Earth is exactly the sort of place we might find one. But to make that point better, we will have to be a bit more specific.

> '...your lesson this time is that you should never lose sight of the alternative.'

The clay-making machine

To Holmes I wrote showing how rapidly and surely I had got down to
the roots of the matter. In reply I had a telegram asking for a description
of Dr. Shlesinger's left ear. Holmes's ideas of humour are strange and
occasionally offensive, so I took no notice of his ill-timed jest...

On the grandest scale, in terms of sheer throughput of materials, the
whole Earth is a continuous crystalliser for **clay minerals**. This is a
chapter about clay minerals: about how they are made and, in some
detail, about what they are like.

Clay is not perhaps generally thought of as a crystalline material,
yet most of it is. If clay seems a rather formless sort of stuff that is
because its crystals are, from our point of view, exceedingly minute:
a mole-hill of clay would have to be magnified to the size of a
mountain before its crystals became visible to the naked eye.

Clay can be roughly defined as a soft rock whose particles are
smaller than a few thousandths of a millimetre across, are rather
insoluble, but can be readily suspended in water. Many kinds of
materials are included in this broad description, and generally
speaking a given piece of clay contains several distinct clay minerals,
each with a characteristic composition of units and/or crystal
structure.

There are two great cycles that drive the clay-making machine. The
first is the water cycle powered by the Sun. Water evaporates from
the sea and other surfaces to form clouds, rain, groundwaters,
streams and rivers that take it back again to the sea. Clays are formed
through the action of this water, through the weathering of hard
rocks such as granite. Such rocks may seem more stable than clays,
but chemically they are not: they slowly dissolve in the waters
streaming over their surfaces and through pores and cracks in them.
Sooner or later the solutions in the porous ground become
supersaturated, they become magic solutions for clays whose tiny
crystals grow there in the ground. Many of these get carried to the
streams and rivers, and may go all the way to the sea, silting into
deep layers on the sea floor. There the conditions will be a bit different
from the conditions under which they originally formed. In particular,

the concentrations of units in solution may no longer correspond to a saturated solution of the clay structure that was. The clay crystals may very well re-dissolve, in that case, to re-form into new kinds of clay minerals. Indeed between weathering and sedimentation – and beyond – there may be several re-crystallisations in different places, each place constituting a kind of continuous crystalliser for a particular set of clays that are being fed by nutrient solutions created by the dissolving of clay and other minerals that are slightly more soluble in the circumstances. This is how the Earth can maintain levels of supersaturation indefinitely within narrow limits in certain regions: the concentrations of input solutions are determined by the solubilities of materials that are just a little bit more soluble than the materials that are crystallising within those regions.

Even for clays in the sediments of the sea floor it is not the end, for there is the second more ponderous cycle that comes into play, the cycle powered by the Earth itself, by the heat inside it that comes from the disintegration of radioactive atoms. This engine pushes sea floors sideways, causing great, slow, catastrophic collisions at the edges of continents so that some of the compacted, by now transformed, clay deposits are more radically transformed at very high temperatures and pressures deep underground. The transformation from hard rocks to clays is being reversed now, because now it is the minerals of hard rocks that are more stable – materials such as the feldspars, micas and quartz of schists and granite. But the pushings and crumplings continue. Sooner or later the re-formed hard rocks re-appear, to be exposed again to the weather, to find themselves, strangely enough, unstable under these mild conditions. They slowly dissolve in the waters streaming over their surfaces and through pores and cracks in them...

Most of the clay minerals that are produced in this way are **layer silicates**, of which there are two main classes corresponding to two main designs for the amazingly thin and beautifully structured layers that their crystals contain.

One of these designs is found in the main component of china clay, a clay mineral called **kaolinite**. Here the layers are about three oxygen atoms thick. Imagine three planes of oxygen atoms stacked on each other (like three layers of oranges in a box) and held together firmly through two intervening planes of much smaller atoms lying in crevices between the oxygens and covalently bonded to them. One of these intervening planes is of aluminium atoms, the other of silicons. There are also (tiny) hydrogen atoms attached to the oxygens of one of the surfaces of the kaolinite layer. (The detailed structure of connections is given in appendix 2.)

You can see that these layers out of which kaolinite crystals are built are quite complicated structures, and they have a 'top side' and a 'bottom side' to them that are different. You can think of one of these layers as being like a carpet with a pile on one side and a backing on the other. Indeed the carpet even has a pattern to it – in effect arrows all pointing one way. This pattern arises from a subtle asymmetry in the arrangement of the aluminium atoms (a feature described in detail in appendix 2).

Now an ideal example of a kaolinite crystal would consist of a stack of many thousands of these figurative carpets lying on top of each other and all the same way up with the pile of one carpet sticking rather firmly to the backing of the carpet above it. Real crystals are often less neat, with great blocks of carpet inverted within the stacks, but we will leave aside such complications for the moment.

The ideal kaolinite crystal would also have another kind of regularity of stacking: it would also have the arrow pattern in the carpet lying the same way throughout a given stack – although again there are complications for real crystals that we will come to in the next chapter.

Among the minor variants of kaolinite there is one called **dickite** that differs only in the way in which the arrow pattern lies between carpets in the stack. In dickite *alternate* carpets have their patterns pointing the same way. This kind of thing, where crystals differ only in the way in which identical layers are stacked on top of each other, is very common in clay minerals. The technical term for it is *polytypism*. Kaolinite and dickite, for example, are said to be **polytypes**. We will be coming back to this too.

Another clay mineral called **halloysite** has kaolinite layers that are much less well stuck together, often with water molecules between. In this mineral the layers may be rolled up, like rolls of carpet, or curled into tiny hollow spheres, or have more complex forms.

In more radical variants of kaolinite the architecture of the layers is maintained while the atoms between the oxygens are partly or wholly changed from aluminium and silicon. There are indeed many variations on the kaolinite theme.

The second of the two main classes of layer silicates is well illustrated by the mica mineral, **muscovite**. It has large well-formed crystals that are easy to study. Not itself a clay mineral – its crystals are far too big for that – muscovite nevertheless has the main architectural features of several important kinds of clay minerals. This is a characteristic layer structure that is a somewhat thicker and more symmetrical version of the kaolinite 'carpet'. (For a detailed comparison see appendix 2.)

The main difference between the two most important kinds of layer silicate is in the way in which their layers stick together. The muscovite layers are quite highly negatively charged and hold together through intervening planes of (positively charged) potassium ions.

These charges within muscovite layers arise from what are called **substitutions** – where, for example, aluminium ions are present in places where more often you find silicon. Each such change from a silicon to an aluminium introduces one negative charge. This is the main kind of substitution in muscovite, responsible for most of the layer charges, but there can be others as well. For example magnesium atoms may be present in places where aluminium is more usual, and that too would give rise to a negative charge at the point where the substitution had taken place. There are indeed rich possibilities for piecemeal substitutions of various sorts in minerals that have mica-type layers. On analysis such minerals usually show that there are many more kinds of metal atoms present than the 'official' silicons, aluminiums, etc.

Among common clay minerals with mica-type layers, the **illites** are similar to muscovite in crystal structure although, being clay minerals, the crystals themselves are very much smaller. Crystals of illite, as one can see under the electron microscope, may be only a few layers thick. As such they are highly flexible lath-like structures and evidently quite tough. Such crystals commonly form in the pores of sandstones at the bottom of the sea.

In **smectite** clays there are fewer negative charges in the layers than in either micas or illites, and so there are fewer metal ions between the layers. These are more often sodium or calcium than potassium and they easily come and go. There are also water molecules between the layers of a smectite and these are more or less mobile too. Water may even come in to push some of the layers in a crystal completely apart. Organic molecules of various sorts can behave similarly – indeed the ability of clays to hold on to organic molecules has been known for a long time and is an important factor in the formation of soils.

Again, as with illites, a smectite crystal can be a very thin, flexible, but quite tough object. It may not look much like the conventional idea of a crystal when seen under an electron microscope: it may look more like a crumpled rag or, sometimes, a folded napkin. A mass of smectite crystals, with layers partly adhering together, partly separated, often has a characteristic 'cellular' appearance, with myriads of compartments interconnected. Typical smectite crystals, like typical illite crystals, are not at all the regular block-like objects that are the more usual idea of what proper crystals should be like. These clay crystals are best described as *membranes*.

Recall that a subtle difference in the mode of stacking of layers on top of each other made the difference between the clay minerals kaolinite and dickite. For mica-type clay minerals there is also a kind of arrow pattern; and again the arrows in layers stacked on top of each other may or may not all point the same way. There are actually no less than six ways in which just two mica-type layers can be put comfortably on top of each other. For each of these ways there would be another six ways of putting on a third layer – and so on. Stacking sequences often turn out to be regularly repetitive, but this is by no means always so. Putting this more technically, there are not only ordered polytypes to be thought about (such as ideal kaolinite and dickite) but also **disordered polytypes** where there may be little if any predictability in the orientation of stacking of different layers on top of each other.

Less subtle, but also very common indeed, is where chemically different kinds of layers are stacked on each other – and again the sequences of stacking may be regular or irregular. For example illites and smectites commonly form such **mixed layer** clays. It is complicated stuff, clay.

How?

There is no doubt that clay structures really do put themselves together, in the sense that they are neither the specially engineered outcome of organisms, nor the products of bizarre geochemical conditions. These are 'zero-tech' materials: they represent ways in which their units, in some sense, want to be – under a wide range of (watery) conditions in surface regions of the Earth. (Indeed it seems that the craving goes beyond the Earth: layer silicate clays are found in some meteorites, and their presence is strongly suspected on the surface of Mars.) What goes on when a tiny clay crystal self-assembles?

When soldiers form up as a platoon; or when soap and water molecules make a bubble; or when sugar molecules crystallise, it is clear in each such case what the units are. In each case disassembled units (soldiers, molecules) simply become packed together in some way without being altered in themselves. By contrast, when a protein or DNA chain is made, the units needed in solution to begin with are not exactly the same as the units in the chain that will be produced by their coming together: pieces of the initial units have to break off as part of the chain-making process (chapter 4 and appendix 1). The formation of clay layers is similar in this respect to the formation of protein or DNA. The units that are in the solutions from which clays crystallise are such things as silicic acic and hydrated metal ions (see appendix 2), and for these units to come together water molecules have to be thrown off. (Only thus can appropriate 'press-studs' be released for a new strongly bonded structure to be built up – as indicated in appendix 2.)

In other crucial respects, though, the making of a clay layer is quite *unlike* the making of a protein or DNA chain, because clay making is a self-assembly, a true crystallisation. For one thing the process is strictly reversible: whether there is a net assembly or disassembly depends, as always with crystallisation processes, on whether the solutions surrounding the crystals are supersaturated or under-saturated. There is no question, then, of the units having to be primed or 'wound-up'. All that is necessary is that the *solution* is 'wound-up' (in the sense of being supersaturated). The Earth seems to be very good at providing solutions that are suitably 'wound-up' for clays – to judge from the vast amounts of clay that are being made all the time. (Just look at a river in spate.)

Another difference stems directly from the fact that crystal growth is a space-filling operation. There is likely to be far more discrimination in packing units to fill space than in linking them together in a wiggly chain. Imagine a three-dimensional crystal of snooker balls – several close-packed planes on top of each other – and then just think of the havoc created by one tennis ball somewhere in the middle. In real crystals the error correction mechanisms that we talked about in the last chapter have little difficulty in detecting mistakes of this sort.

Finally, as a clay crystal grows, however reversible the bonding may be at the surface and in contact with water, the bonds that get buried inside and away from the water no longer have the means (or the elbow room) to effect changes at all easily at ordinary temperatures. They become faithful then, like concubines, through lack of opportunities.

> I shook my head. 'Surely, Holmes, this is a little far-fetched,' said I.
> He had refilled his pipe and resumed his seat, taking no notice of my comment.
> 'The practical application of what I have said is very close to the problem which I am investigating. It is a tangled skein, you understand, and I am looking for a loose end.'

12

Gene-1

'I'll tell you one thing which may help you in the case', he continued, turning to the two detectives. 'There has been murder done, and the murderer was a man. He was more than six feet high, was in the prime of life, had small feet for his height, wore coarse, square-toed boots and smoked a Trichinopoly cigar. He came here with his victim in a four-wheeled cab, which was drawn by a horse with three old shoes and one new one on his off fore-leg. In all probability the murderer had a florid face, and the fingernails of his right hand were remarkably long. These are only a few indications, but they may assist you.'

I'll tell you one thing that should help us above all in the case of the origin of life on the Earth. The first organisms had genes in them. These genes were, in all probability, microcrystalline, inorganic and mineral. They crystallised continuously from water solutions that were being maintained slightly supersaturated, over long periods of time, somewhere near the surface of the Earth. These are only a few indications, but they should assist us.

There are more indications when we come to think in more detail about the nature of genetic materials. Surely a genetic material, even a complete beginner, must be singular stuff, if it can hold and replicate mutable information that can affect its own survival... As I said before a crystal gene is not going to be just *any* kind of crystal.

That a crystal gene has to be able to hold information tells us immediately that it cannot have a completely regular structure. This is a general point. If you were to turn the page of this book, for example, and were to find only a regular pattern of letters... abcabcabcabcabcabc...line after line, you would know that little was being said – even if you suspected that I had suddenly decided to write in code. On the other hand if you found a random-looking sequence then it might be carrying information for all you would know. What if you found only ' A 's scattered like confetti at all angles over the page? Well, again, that might be carrying information – provided of course the letters were not actually confetti letters that could move about too easily, and provided that they had not actually been scattered at random. One could imagine that the multitude of orientations and positions of ' A 's on the page represented an encoded message of some sort.

Printed pages in books are typical of informational structures in having at the same time regular and irregular features. That the letters are arranged in rows, are of much the same size, and are all one way up: these are regularities. There are more subtle regularities in the letter sequences themselves since these are more or less constrained by rules of grammar etc.; but they are not completely so constrained and it is precisely to that extent that they can convey information. In general we can say of any informational structure that the more random it *seems* the more information it might contain.

The need for fixed irregular features, then, is clear enough. But what is the point of all that regularity that is so characteristic of informational structures from printed pages to DNA molecules?

Surely it is that the regularities make the information easier to handle – pages are easier to print, DNA molecules are easier to replicate, both are easier to read.

Indeed DNA has only quite a limited information capacity with its relatively small structural variability – a mere one-in-four choice for every sixty atoms or so. This might seem odd for such an important information store until you realise that sheer information capacity is easily achieved. It is easy to be irregular. The difficult bit for a molecule is to have specific irregularities that are replicable and meaningful. So if regularity helps here – if for example replication is easier for DNA because of its uniform connector pieces (see p. 23) – then ask no more: the regularity is helping to overcome the really difficult part of the problem.

All this suggests that when thinking about the nature of primitive genes we should perhaps be thinking of structures with still more regularity in them than DNA molecules – where the irregular features are, perhaps, but occasional modulations to an underlying repeating pattern.

That brings us back to crystals, because just about all crystals are like that: they have an underlying crystal structure on which is superimposed a **defect structure**. The first is, broadly speaking, a characteristic of the material; the second is characteristic of an individual crystal and consists of various kinds of irregularity. Because of their fixed defect structures, you can be pretty sure that no two quartz sand grains on the seashores of the Earth are exactly the same. So let us try to think about defects in a crystal of the sort that might replicate as the crystal grows.

A perfect crystal is necessarily a fiction, since it is an infinite array of units in three dimensions. The finite sizes and particular shapes of real crystals are thus already defects – however grudging a description that may seem. To be sure, shapes and sizes are characteristic of individual crystals and to this extent they could represent specific information. The shapes and sizes of the tiny crystals of clays are important in determining the large-scale properties of these materials: so shapes and sizes might indeed have some kind of rather direct 'meaning'. But the 'meaning' aspect of genetic information will be the main theme of the next chapter; in this chapter we will concentrate on the other key aspect – replicability.

So could shape and size be a replicable property of a crystal? It is hard to see how an arbitrary three-dimensional shape and size could be replicated through crystal growth processes; but a two-dimensional shape and size – a cross-section – might very well be. All that would be required would be that the crystals should grow in one direction only and break up (only) across that direction:

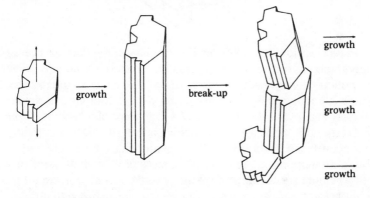

It is quite common for crystals to grow much more rapidly in certain directions, developing needle-like or columnar forms. And it is quite common for crystals to break more easily in some directions than others: so it is not, presumably, a freak combination of properties that is required. Indeed that common clay mineral, kaolinite, quite often has columnar crystals that seem to grow and break up in the way required. Such crystals are often complicatedly fluted and grooved suggesting that a fair amount of 'shape-and-size information' is indeed replicated through their growth and break-up.

There may be more here still than meets the eye, because a complicated fluting or grooving on the surface of a crystal is an indication of internal defects of a sort in which different regions within the crystal are somehow misoriented with each other. This is a very common type of crystal defect, especially in minerals. The simplest form of it is called **twinning**. A twinned crystal is like a badly fitted carpet with the pattern on different sections plausibly but incorrectly matched up. The sort of mismatching suggested by the grooving of these crystals would be one in which each layer making up the columnar stack had the same mosaic pattern of misorientations:

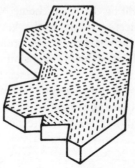

These clues that we have been talking about come from the overall appearance of certain kinds of kaolinite crystals – what are called 'vermiforms' – especially as seen under the electron microscope. There is another line of enquiry, using X-rays, that gives a similar picture. Recall that the structure of a little piece of a kaolinite layer is directional – it has a kind of 'arrow pattern' in it. More explicitly this pattern can lie in any of three orientations. (The details can be found in appendix 2.) Well, in a big piece of a kaolinite layer in a real kaolinite crystal all three of the possible orientations are there: a given layer is in fact made up of a mosaic of patches with different orientations, like the above picture.

Another thing that is known from X-ray studies, and was mentioned in the last chapter, is that in kaolinite crystals a given orientation for the 'arrow' in one layer tends to be repeated in all the layers above and below. We have then a plausible detailed explanation not only for the complicated cross-sections of kaolinite 'vermiforms', but also for why a particular complexity appears to be retained over many thousands of layers: it is because a new layer, as it grows, copies a defect pattern, a particular mosaic of orientations in the layer underneath it.

That is still conjecture: but it is a conjecture based on fairly normal kinds of crystal growth processes. But, you might ask, what about those much talked of error correction mechanisms that are supposed to operate through crystal growth processes and keep them on the rails? How do they put up with all these mismatchings? Are these not exactly the kinds of things that error correction mechanisms are supposed to eliminate?

In this particular case, because of the unsymmetrical nature of the kaolinite layer structure, there is reason to believe that a mosaic arrangement of some sort would actually be more stable than an arrangement with all the arrows one way. Even so, one might have expected that some regular, ideal, most stable mosaic pattern would be produced as a result of error correction mechanisms. Put most generally, and as a general criticism of the whole idea of crystal genes, it might seem that the very trial-and-error processes that are supposed to give fidelity of replication only apply to *regularities* in crystals and not to the *ir*regularities that alone can hold information. It might seem that from the point of view of a growing crystal any 'information' is an error to be done away with. (Have I really brought you all this way just to tell you that replication of information through crystal growth is, after all, impossible?)

Yet this is surely not a fatal objection, because in practice we know perfectly well that real crystals grown from real solutions do contain myriads of defects: evidently the error correction mechanisms do not always work. And, from our discussions in chapter 10, we know why. It is because solutions are hardly ever at low enough levels of supersaturation for the error correction mechanism to be able to detect those kinds of defects that have only a small effect on stability.

Let us concede that an information-containing structure of any sort whatever must be 'metastable', that is to say a fixed arrangement but precarious to the extent that it is not the most perfectly stable arrangement possible (there can only be one perfect anything): but – and here is the let-out – there is no relationship whatever between the amount of information that structures can carry and how unstable these structures are. It all depends on the particulars of how the information in them is being carried. A structure that is only a *minute* amount less stable than the ideal may very well hold vast amounts of information. The arrangements of letters in this book, for example, or the sequences of units in your DNA molecules: these hardly affect at all the stabilities of the structures that hold them.

So you see, as is so often the case with 'fatal objections' of the theoretical sort, this objection to the crystal gene has served simply to clarify a requirement: the information-containing defect structure should only have a small effect on stability.

How might we decide what sorts of defect structure are like this?

Look at Nature: find what kinds of crystal defects are common in minerals. You can take it that these will be the sorts that have little effect on the stabilities of the structures that hold them. Twinning is one such; we will consider a few other sorts shortly.

Of course this need for 'marginal metastability' is only the start. The next requirement I can only call cheeky. Not only must the defect structure fail to be eliminated by crystal growth processes, it must get itself *repeated* by these processes.

This is not so difficult to imagine really. It is often easier to repeat an irregularity than to eliminate it. For example, in laying tiles, one tile placed too far to the left may lead to a whole row similarly displaced.

Now imagine a twinned crystal with a mosaic pattern on its surface made up of different orientations of a directional crystal structure – as in the picture on p. 90 – but let us assume for simplicity that a uniform orientation really is the most stable arrangement. The irregularity could be eliminated in the very next layer if only this time the units would all line up the same way. We can see that this would be the way to get rid of the imperfection and arrive eventually at a more stable kind of crystal, but the units could have no such insight. For them this first move towards perfection might be very far from a stabilisation, as two thirds of some set of units, however nicely matched-up with their sideways neighbours, would be mismatched with the units in the layer below them. If the zones were extensive it could very easily be the case that the (then only occasional) improvement in sideways matching would be nowhere nearly balanced by the disastrous number of vertical mismatches. 'If you can't beat 'em, join 'em' is one of the deeper principles of crystal growth.

And the lesson from all this? If you want accuracy of copying above all else, sacrifice information capacity, have extensive zones – write big.

So far we have been talking about physical defect structures. There are also chemical defects, where there is some irregularity in the kinds of atoms present in a given crystal structure. In this case the 'carpet' is matched-up as far as directionality of the pattern is concerned, but some of the roses in the pattern are cornflowers instead. You could of course write messages in a carpet that was like this. In a similar way substitutions of metal atoms in mica-type clays of the sorts described in the last chapter provide an abundance of possibilities for information storage.

Those substitutions that generate negative charges within the silicate layers of clays are particularly interesting. Could a specific arrangement of negative charges in one layer be inherited by a new layer growing on top of it?

It seems quite possible. Recall that the negatively charged silicate layers have 'loose' positively charged metal ions between them: these positive ions hold the layers together. So one could imagine a negative charge in the top layer of a stack attracting a 'loose' positive ion, which then encourages another negative charge to be located in that position in the next layer as it forms. In support of this notion one may point to a number of known layer silicate structures in which centres of positive charge lie precisely between centres of negative charge in the layers above and below. That kind of thing makes electrical sense, but one cannot be sure from these cases whether an actual inheritance of particular charge arrangements was involved during the growth of their crystals.

To try to find out that sort of thing calls for a study of the crystal growth processes themselves. In 1981 Armin Weiss of the University of Munich reported the results of laboratory studies on the growth of smectite crystals, where new layers were made to grow *between* layers in pre-existing crystals. The new layers then had a similar charge density to the layers between which they had grown. Furthermore this kind of inheritance could be maintained for over 20 'generations'. Although the details have yet to be published it is clear that the conditions required for these interesting experiments were highly ingenious and contrived.

We can say of the kind of copying process that Weiss describes that it puts a similar amount of ink on the copy as was there on the master. This is at least a *sine qua non* for any direct replication process and, given this, one might expect that at least large-scale features of charge pattern would be replicated also. We do not know, though, if this is the case: we do not know, even, how sharp such printing *could* be. If it could be really sharp with one-for-one, charge-by-charge copying between layers, then the replicable information capacity would be quite comparable to that of DNA. As a piece of sheer engineering it is a fascinating thought, with all kinds of consequences: but perhaps we should be less ambitious in our thinking about truly primitive genetic materials. Fidelity is what mattered for gene-1, and that is not the same thing as fineness of grain. Errors, as we saw, may be less likely with big print – and operating conditions are likely to be less exacting.

Another potentially helpful form of extravagance, to improve the security of the information and the fidelity of its copying, would be through redundancy (repetitiveness) of the information: that is to say through the crystal consisting of an array of many copies of the message being printed. The kaolinite model for a crystal gene which we considered earlier is like that: it is like a book with exactly the same thing printed on each page – rather a disappointing book. But with a set-up like this, although the specific patterning is in two dimensions, this patterning is nevertheless firmly part of a three-dimensional crystal. It is much easier to imagine accurate replication, in that case, based on the error correction mechanisms of crystal growth. Single silicate layers would, I think, be too flexible to get themselves easily copied: it would be much better to have the stable surface of a fairly thick, rigid crystal for the replication of a pattern on that surface – a pattern all the more secure because it runs all the way through the crystal. (These would be 'Brighton Rock genes'.)

And the whole business could be more casual too, in a number of ways. Because there are so many copies, what does it matter if a few layers are dissolved away accidentally? And when it comes to the later part of the replication process, the break-up, what does it matter exactly where the pack of cards is cut when all the cards are the same anyway? Note that single 'cards', i.e. single layers, need never be produced: the break-up never has to go that far. This is just as well, perhaps, for another reason: it is difficult to find conditions that are at the same time compatible with separation and with crystallisation of the silicate layers. How would partly made layers stay in place? How would they know when to separate? Great stacks of layers would have no such problems, they would simply break mechanically when they got too big.

Let us be still more general for a moment before moving in on another kind of crystal gene. Let us think a bit more about dimensions.

Informally we may say that a sheet of paper is a two-dimensional thing, a piece of string a one-dimensional thing. This is not quite true of course, because all real objects are in three dimensions. Yet when we draw a picture on a piece of paper we are only 'using' two dimensions, and when we crumple a piece of paper we are making use of the fact that there is a spare dimension to be crumpled into. Similarly, unless you are a very small insect indeed, there are really only two ways to go on a piece of string – although (thanks this time to two spare dimensions) many, many ways to fold and tangle it. We may call these roughly defined properties of sheetiness or stringiness physical dimensionalities.

Informational dimensionality is a more abstract idea, but it is usually clear enough. Writing holds information in one dimension, a blue-print in two dimensions, while a model of a proposed building would be holding information in three dimensions.

So when we say that a DNA molecule is a one-dimensional information store we mean it in two distinct ways: first, because a DNA molecule is physically like a piece of string, and second because the information in DNA is held in one dimension – it is a sequence. Other informational structures are less neat: for example this two-dimensional page has a one-dimensional message on it; a three-dimensional protein molecule may also have only sequence information in it.

And then of course the crystal genes that we have just been talking about are three-dimensional objects containing two-dimensional information. The whole idea is that a Proper Crystal Gene should be firmly a three-dimensional object, its units completely filling space so that packing requirements will provide the discrimination needed for the replication-through-crystal-growth mechanism. On the other hand the information itself must not be three-dimensional if it is to be at all easily replicated: for a simple copying process there should be at least one spare dimension for it to be replicated into.

It should be clear now that there is another kind of Proper Crystal Gene to be thought about. Information could be in one dimension replicating into the other two.

In place of that rather disappointing book in which everything was written on the first page, and then all the other pages were just the same, think now of a really tedious book in which virtually nothing is written on *any* page – where there are only two or a few simple standard pages, and where the information is simply in the order in which these pages are stacked. You might think of just two kinds of pages: one kind has nothing but 'a's neatly typed all over it while the other kind has only 'b's. The message would be in the stacking direction as, say, aababbaabbbbbabbbba. If you remember (p. 84), mixed layer clays consist of different kinds of layers which may often be superimposed in a more or less irregular manner.

Indeed complicated stacking sequences are quite possible even with only *one* kind of layer. Recall again (p. 82) that the difference between the ordered polytypes kaolinite and dickite is a difference in the way in which identical layers are superimposed. (The pattern always points one way for kaolinite, but alternates first one way then another for dickite.) Imagine pages with only a's printed on them but with the letters different ways up between the pages of a stack: say aaaaɐɐɐaɐaɐɐaɐaɐaɐaaaɐɐɐaɐaɐɐ. A disordered sequence like this could hold information too, of course; and as remarked earlier, disordered polytypes are common, and occur often in clay minerals, particularly in mica-type clay minerals.

Nor is the amount of information that could be held in this austere way by any means trivial. For example with six different ways of putting mica-type layers on top of each other a stack of 140 such layers could in principle hold as much information as a record of 140 throws of a dice – and we saw (p. 47) how seriously you would have to take that kind of complexity of message.

If you are still thinking that stacking sequences would be an excessively wasteful means of information storage then just remember DNA. For 'page' read 'nucleotide letter pair' and you will see that it is much the same idea. An imaginary (*very* small) insect will see that the DNA string is a double flex within which there is indeed a stack of tiny layers on top of each other, the information being simply in the sequence of stacking (four kinds of) these layers. In a Proper Crystal Gene with one-dimensional information, the layers would be hugely extended sideways with no need for a flex, then, to keep them in place, and a more casual style of replication.

As with the previous kind of crystal gene, replication would involve both crystal growth and crystal break-up. As before there would be rules for the directions in which these processes took place. This time, though, the rules would be the opposite way: here growth should be exclusively sideways, never layer-on-layer; and break-up must not separate the layers but cut right through them. Imagine a book that grows by the pages getting bigger and bigger (but which never gets any thicker) and which is then guillotined into smaller books whose pages continue to grow bigger and bigger...

It is curious that in spite of such different requirements layer silicate clays can again give us models: some mixed layer clay minerals have appropriately shaped crystals. Illite–smectite, for example, commonly shows extensive lath- and sheet-like forms suggesting a strong preference for growth sideways. These seaweed-like structures, tough and flexible though they are, are liable to be torn. What a crude way of completing a replication cycle – for bits just to get torn off from time to time! Yet if it would work it would do for gene-1 who, you can be sure, was not at all stylish. Just because the information is so repetitive the break-up can be casual: the guillotine is unexpectedly not required.

But let us not insist too specifically on layer silicates for gene-1 (or genes-1). We homed in on these materials in the first place because so much of the Earth's clay is of this sort of stuff: but there are many other kinds of minerals that form tiny crystals from water solutions – clay minerals in a more general sense – and such features as twinning, stacking errors, cation substitutions, growth in preferred directions, or break-up along preferred planes are common in crystals of various sorts and found in minerals in various combinations.

We have, as it were, identified the organisation responsible for that 'crime against common-sense', the origin of life. And it is true that the proposition that our ultimate ancestors were mineral crystals was not widely anticipated. We even have some individuals on the list of suspects. But we are still some way off making an arrest. Has this extra complication that we now see at the beginning of evolution made the whole question of the origin of life more remote than ever?

I do not think so.

> '...the very point which appears to complicate a case is, when duly considered and scientifically handled, the one which is most likely to elucidate it.'

13

Evolving by direct action

...I saw by the Inspector's face that his attention had been keenly aroused.
'You consider that to be important?' he asked.
'Exceedingly so.'
'Is there any point to which you would wish to draw my attention?'
'To the curious incident of the dog in the night-time.'
'The dog did nothing in the night-time.'
'That was the curious incident', remarked Sherlock Holmes.

It was once thought that organisms quite often arose directly, through transformations of matter that did not require the reproduction of previously existing organisms. Such **spontaneous generation** was thought to be commonplace: for example frogs were thought to arise *de novo* from mud, and maggots from decaying meat.

I now wish to draw your attention to the curious incidence of spontaneous generation as we now see it. There does not seem to be *any*.

Is that not rather odd? Once upon a time an evolutionary process started up, but there are seemingly no brand new beginnings any more. The tree of life flourishes: but apparently there is only one tree with no sign, even, of any recent saplings on the same ground. *Why are there no signs of present or recent spontaneous generation?*

The first answer that may come to mind is that organisms are just much too complicated. One can easily see the absurdity of spontaneous frogs or maggots, now that the 'high-tech' nature of such forms of life has become so apparent. Similarly for bacteria – *E. coli* is not really a simple thing either. And since bacteria are nevertheless among the simplest free-living organisms that we know of, why then, any spontaneous generation may now seem absurd.

But wait a minute: that line of thought by-passes the argument. The point is that we might have expected, on the assumptions that we have been making, that there should be *some* really simple kinds of organisms around us. After all we have been assuming (chapter 1) that the first organisms really did appear spontaneously on the Earth (with no miracles, freak events, frame-ups or alien infections). In that case there should at least *have been* organisms of a kind that could generate spontaneously. Remembering the limitations of pure chance as an engineer (chapter 6) such organisms would have been, chemically, child's play. Presumably the laws of Nature were not any different 4 billion years ago. So why are there not still *such* organisms to be found – self-starters, what I will call **primary organisms**? Why do we not know all about them?

Is it because the general conditions on the Earth have changed? Is it that the conditions for making and sustaining the first sparks of life are no longer suitable? This is a standard answer. But we have seen how insufficient this kind of answer is. Just having general conditions right would not have been nearly enough. Even with the most favourable general conditions imaginable, key molecules of even quite modest complexity – nucleotides, for example – could not have been made with anything like the necessary competence. To make molecules of that sort requires an intricate interference, an elaborate manufacturing procedure that only the prolonged operation of natural selection (or an experienced research chemist) could reasonably be expected to generate.

Anyway *we* should be able to do much better than just create suitable general conditions. We *can* set up manufacturing procedures and automatic machinery to carry them through. We can load the dice – we can contrive situations and devices that could by no stretch of the imagination have arisen *de novo* by chance. So if Old Fumble Fingers, pure chance, did indeed put together the first evolvers under the rules of chemistry and physics, why, for goodness sake, can't *we*?

If, as it would seem, the difficulty is not technical, it must be a difficulty of knowing which area of possibilities to explore, of seeing the appropriate design approach. Perhaps, in line with the whole drift of this book, research has been concentrated on the wrong materials and on the wrong natural phenomena? (Are new sparks too gentle to be easily seen, now, against the more general blaze?)

But if, as I have been saying, the key materials for primary organisms are inorganic crystals (so that the topic of the origin of life on the Earth is a branch of mineralogy) then the opening question of this chapter becomes rather sharp – all the more so now that the pre-vital Earth is being seen as having been more 'normal' than was thought previously (p. 6).

Nor is there a let-out in the explanation (given on p. 42) for why the single common ancestor for all life on the Earth should have been so highly evolved. That kind of explanation pre-supposes the absence of recent spontaneous generation. It says that *if* there was only one tree then almost any arbitrary combination of branchings and prunings would have led to a late last common ancestor. But with spontaneous generation we should expect there to have been not one tree but quite a little forest of trees of different ages, and in that case no single common ancestor.

Of course a random-branching-and-pruning picture of the growth of evolutionary trees is too simple. There would have been other factors at work. One of these was pointed out by Darwin: organisms no longer originate *de novo* because any brand new forms of life would be promptly eaten by evolved forms. No doubt this is so. But how promptly? If the very first forms had been inorganic crystal genes, they would have been unappetising to any organisms of the modern type. Darwin's point explains, perhaps, why there are no obvious well-developed saplings around the tree of life. But on the crystal gene idea should there not be myriads of very small saplings, near-starter organisms that have not got far enough to interact with modern organisms *because not yet sufficiently like them?*

And of course that could be a reason why primary organisms have not yet been recognised: because they are made of the 'wrong' materials, and because their appearance is not 'life-like'. I think it is quite possible that primary organisms are indeed all around us.

By chapter 9 we had arrived at the idea that the first organisms would most probably have been made from substantially different materials from today's biochemical materials, and we touched on the notion that these organisms might have been 'naked genes'. Crystals then seemed to be the best bet for such genes.

But would not the term *organism* be altogether too pretentious for, well, just a mass of tiny crystals? One might be inclined to insist on something more interesting before using the word organism to describe it.

That would be sheer prejudice. We are never going to find (or make) primary organisms if we have too high-flown ideas about what they should be like. They will be all potential with little or no achievement. Of course they are going to be rather boring, poor things. If our ultimate ancestor was indeed a product of the Earth then similar things that we might find now should be similarly *mineral* first and foremost.

For example, clay crystals growing in the pores within a piece of sandstone might very well turn out to be primary organisms. Clay crystals growing under such conditions have, often, distinctive and elaborate forms – such as the grooved kaolinite vermiforms that were described in the last chapter – and it is not too difficult to imagine circumstances in which simply the shapes and sizes of crystals could have a bearing on their ability to grow quickly, or break up in the right way, or stay in the right place, or survive difficult conditions – or otherwise be a success. Provided some aspect of the shape and size (e.g. a cross-section) is replicable, with occasional errors in the replication, then that aspect will be subject to natural selection: it will tend to become optimised in a way that is formally just the same as the way in which the parts of plants and animals become optimised through natural selection. The practical difference here between crystal genes and, say, trees or giraffes would be that for crystal genes shapes and sizes are so much more directly specified by the genetic information. Indeed for real beginners the messages in crystal genes may simply *be* aspects of shape and size. Naked genes are direct-acting genes and enormously simpler participants in evolution as a result (pp. 66–7).

The shapes and sizes of clay crystals can greatly affect the porosity of a sandstone that contains them. This allows one to imagine a process of natural selection operating at a very simple level. Imagine a piece of sandstone that has initially two different crystal genes in it. Each of these soon makes a little zone containing thousands of tiny crystals, all the crystals in one zone having similar features of shape and size and hence giving a characteristic porosity to that zone. The crystals in one of these zones are so shaped that they clog the pores completely. The flow of nutrient (supersaturated) solutions stops in that zone, the flow now being channelled elsewhere. These crystals stop growing. The crystals in the other zone have a different characteristic shape which is being replicated – a rather spindly shape, perhaps, that allows the crystals to grow without completely filling the space in which they are growing, so that solutions can still flow through to continue the supply of nutrients. It is this second form which thus tends to spread to fill the sandstone with its characteristic mass of loosely woven crystals. Bits of this fabric break off, perhaps, to carry the secret of how to grow effectively to other pieces of sandstone, which then become infected. Inevitable mistakes in the replication processes would ensure that there was always some spread of types. I dare say the detailed shapes would not have much effect on performance: but real cloggers would always drop out of the race, as would, for example, crystals that were too small and hence too easily washed out of the growth region altogether... In the end you would expect, not always one exact shape within the sandstones of a given region, but a general style or perhaps a few styles if there was a spread of types of sandstone, or a spread of flow rates, or more generally a spread of niches calling for somewhat different optimal shape characteristics.

Now imagine a slightly more complicated situation. Let us suppose that sometimes the flow of solutions through the sandstone becomes rather fast and that under these circumstances the waters are generally undersaturated having not had enough time to dissolve the hard rocks through which they had been flowing before reaching the sandstone. Now there is a new problem for the clay crystals: to avoid being re-dissolved when this happens. One idea would be to become impervious under these circumstances. It sounds like a tricky combination of properties, but when the flow becomes too fast and turbulent the laths fluff up into a tangle that cuts down the local flow rate.

This is but another plausible (I hope) story, strictly for illustration: to show how easy it is to imagine particular circumstances that could provide selection pressures in favour of crystals with some fairly simple characteristics of the sort that might be replicable. But I must confess to having had illite in mind for that last bit. These clays grow in vast amounts in marine sandstones as thin, flexible, seaweed-like structures having odd effects on the porosities of the sandstones that contain them. (And a great nuisance they are too for under-sea oil recovery. They tend to clog up oil-bearing sandstones.)

These illites, if they can be described as organisms at all, would certainly have to be put low on the evolutionary scale: as crystal genes with one-dimensional information their information capacity would be very limited since they only have a few – often only three or four – layers in them. (If you remember, in crystal genes of this sort the information would be held as a sequence of layers.) But that some putative naked-crystal-gene could hold only very little information would be less important than that this information was (1) replicable, (2) adaptive (that is to say affected the chances of success of the gene material holding it), and (3) could be further elaborated in the future.

Well, (1) and (2) are a good bit more than conceivable, as I have been trying to indicate in this chapter and the last; and (3) presents no problem in principle, even for humble illites. A hundred layers instead of four would provide a very substantial information capacity, far beyond the reasonable expectations of dice-throwers. So, partly, it would be a question of whether in some circumstances thicker crystals might be better than very thin ones: and then it would be a question of whether particular features of the stacking of layers made any difference to their properties – properties such as flexibility, modes of break-up, or rates of sideways growth. Again this would be more than conceivable...

We are already well beyond the stage of even trying to work out what *must* have happened. What *does* happen in evolution depends so much on particular circumstances that the course of evolution over the long term is about as predictable as the meandering form of a river or the exact shapes of tomorrow's clouds: one can only illustrate possibilities and indicate general expectations. Of course we cannot know which particular circumstances really mattered for the very beginnings of evolution. But we do know that the real world is full of particulars – structures, events, situations – and that such can be of critical importance in determining the course of evolution. Organisms do not just evolve, they are driven to it by their surroundings, by all sorts of detailed threats and opportunities.

Here is an example of the sort of thing I mean. At some stage in the evolution of plants the woody stem was discovered. This was tied in with the discovery of a strengthening material called lignin – quite a complex substance that can be manufactured in a few steps from one of the protein amino acids. Plants that could make lignin could grow tall and steal the sunlight from others. The characteristic shape and size of a tree, with its tall, rather open branching structure, no doubt depended on the discovery of a suitable strengthening material. But the tree was to create new threats and opportunities for other organisms. Take a walk in the woods and see. Or think about our ape-like ancestors. It is not so much that they chose to live in the trees, rather they were driven into existence by, among other things, the existence of trees: that sort of animal was partly *caused* by trees. So, then, were we. It is at least plausible that our eyes point forwards mainly because our acrobatic ancestors needed to be good at judging short-range distances; that our excellent hands were designed mainly to grip branches thick enough to support body weight; that our good sense of balance that allows us to learn and enjoy such extraordinary feats as skiing or cycling also harks back to a more necessarily acrobatic way of life. (Who would have thought that lignin had anything to do with cycling? Yet I think it is true to say that if lignin had not been invented then man, that package of curious abilities, would not *exist*.)

Such are the unforeseeable consequences of evolution – even of those parts of the process that are comparatively near to us. How can we say anything about the very earliest stages? We certainly cannot say exactly what happened, because all sorts of details of circumstance are lost. But principles are understandable: we should be able to understand the kinds of situation that would encourage evolution, and the kinds of direction that evolution might be expected to take.

In the search, here and now, for primary organisms, it is particularly important to think about the kinds of situation that would actively encourage evolution – just because primary organisms, unlike other kinds, can exist in unevolved states which may be difficult to recognise. If primary organisms are to reveal themselves, they should be exposed to threats and opportunities. We might look for evolved primary organisms in places that are adjacent to places in which clays can grow easily: where only clays with rather special properties will be able to survive. One can imagine primary organisms starting (all the time and all over the place) in easy regions and then (sometimes) evolving into more difficult regions.

Why should they do that? Why should they not just stay in the easy regions? Why should they not just stay unevolved?

Most of them would, I dare say. The question is why some should take a more difficult path. Well, one might as well ask why some animals came out of the sea onto dry land (a far more difficult place to live); or why the ancestors of birds took to the air; why our own ancestors took to the trees; or why we will sooner or later colonise Mars. It is not for the sake of an easy life exactly, as a great deal of difficult technology has to be developed in each such case. But, given the technology, there are then new opportunities for the organisms that have it. It is clearly an important direction of evolution that organisms gradually come to occupy more and more difficult niches – that they aquire the means to survive where others cannot. This is not a question of innate ambition: it just happens as a consequence of natural selection operating in a complicated and heterogeneous world.

Another principle, and a trend that goes hand in hand with increasing control of the environment, is from more direct to less direct means of genetic action. In chapter 7 we considered how the consequences of this are built into the common control structure of all the organisms that we know about. Indirect action makes a lot of sense. By acting so indirectly – through RNA, proteins, cells, higher-order structures – those dry DNA messages develop rich and varied meanings: a dull-looking score is orchestrated and performed.

Now primary crystal genes evolved by direct action to begin with, but would they not too have moved towards a more indirect control? A genetic material has quite enough on its hands holding and accurately replicating its messages. It is unreasonable to expect it to be the last word as, say, a membrane material or a catalyst *as well*. To have one material, DNA, as the information replicating specialist, and a quite different material, protein, as the Jack of most other trades, is wholly comprehensible as an eventual outcome of evolution. Many, many more properties can be controlled by working indirectly like this. For evolving primary organisms the tactics could not have been through protein or anything like that. But surely the general strategy would have been similar, because the logic is similar for organisms at any stage of evolution. Indirect operation is always likely to provide a route to a greater variety of means of control, and in a complicated and heterogeneous world that means a greater variety of places to thrive in.

Trying hard to remember that the first organisms had no sight of the DNA-based machinery that lay in the distant future, we might now ask in what sorts of ways gene-1 might have extended its control. What sorts of separate phenotype structures could it most easily contrive, and for what purposes?

Different kinds of clay mineral crystals may grow in collaboration, one kind affecting the conditions for others. This may happen through new crystals forming on the surfaces of crystals that are already there – a kind of seeding. Or it may happen more indirectly through one clay altering general conditions, such as flow rate or local acidity, which then favours the formation of other clays that might not otherwise have formed. In so far as replicable features (e.g. shapes, sizes, surface patterns) of crystal gene clays could affect the formation of other non-gene clays, and in so far as the non-gene clays might be helpful to the gene clays, then the non-gene clays would be properly described as phenotypes of the gene-clays.

Picture these now somewhat evolved organisms as consisting of masses of crystal genes embedded in a watery matrix of other clay or clay-like material. It is not difficult to imagine uses for such a matrix material. It might provide mechanical protection against damage (growing crystal genes must break, if you remember, but only in the right way); or it might act as a glue, holding the genes in the right place. Or the matrix might provide protection against the effects of fluctuations in concentrations of nutrient solution: if the helper materials were to grow and dissolve more quickly than the gene materials, then the helper materials would have such a stabilising effect on the waters in their surroundings. Or again the matrix material might tend to hold on to metal ions that would interfere with the growth of the crystal genes...

It is almost too easy to imagine possible uses for phenotype structures – because the specification for an effective phenotype is so sloppy. A phenotype has to make life easier or less dangerous for the genes that (in part) brought it into existence. There are no rules laid down as to how this should be done.

We are now moving away from the relative security of unevolved primitive genes where possibilities are constrained by more or less well-known characteristics of materials – of crystals and molecules. We are moving away from direct-acting genes to a new playground, to convolutions of indirect genetic control that seem to be without limit.

> 'Ah! my dear Watson, there we come into those realms of conjecture, where the most logical mind may be at fault.'

14

Takeover

'...there should be no combination of events for which the wit of man cannot conceive an explanation. Simply as a mental exercise, without any assertion that it is true, let me indicate a possible line of thought. It is, I admit, mere imagination; but how often is imagination the mother of truth?'

If there is anything at all in the story that I have been telling so far in this book it is that the problem of the origin of our biochemical system, with its various molecular components, is to be distinguished from the real problem of the origin of life. Evolution did not start with the organic molecules that have now become universal to life: indeed I doubt whether the first organisms, even the first evolved organisms, had any organic molecules in them at all.

Hence the delay in bringing organic molecules into the story. But what was the bridge between evolving mineral organisms and the altogether different form of life that now dominates the Earth? The overall character of this bridge has already been indicated in chapter 8. There was a takeover: the first organisms, as they evolved, created within themselves the conditions under which 'high-tech' genetic systems could appear, then operate with increasing competence – and then take over. Primary organisms were displaced by **secondary organisms**, that is to say organisms of a kind that would have been quite unable to generate spontaneously.

Let us now try to sketch-in the required bridge in more detail – simply as a mental exercise, without any assertion that the details are true. We can proceed through four questions:

How did organic molecules come in – and *why?*
How did organic molecules win – and *why?*

On the first of the 'How?' questions, there was that clue, if you remember, in the Chinese box, which suggested that our biochemical supply structure was built up from carbon dioxide, and *that* suggests that it was through *photosynthesis* that the first organic molecules were acquired by mineral organisms.

This at least would be one constant factor through evolution. Photosynthesis is still by far the most important way in which carbon atoms are incorporated into organisms. It is the unique ability of plants to use the energy of sunlight to make bigger molecules, indeed to make most of their substance, out of two of the most readily available materials – water, and carbon dioxide from the air.

Not that the primitive process could have been anything like as sophisticated as that in modern plants. In green leaves there are tiny machines, each with a thin membrane that contains crystal-like arrays of chlorophyll molecules. These arrays catch packages of sunlight, rather as an aerial catches radio waves, creating disturbances which culminate in the separation of electric charges and their movement to opposite sides of the membrane. The positive charges go one way to act on water (making oxygen gas as a by-product) while the negative charges, electrons, go the other way to help to make sugar molecules from carbon dioxide.

It is quite a complicated process, but there are minerals that mimic the effect of it to a limited extent. Under ultraviolet light some simple iron salts dissolved in water can 'fix' carbon dioxide into small organic molecules such as formic acid. Some crystalline minerals also act in this way.

In another key process, nitrogen atoms are built into biochemicals through taking apart nitrogen molecules in the air. This is not easy because nitrogen molecules, like carbon dioxide molecules, are rather stable. Much energy is needed for *nitrogen fixation*, and it is only achieved by certain bacteria. Again it is a complicated process within organisms, but again there are minerals that can do something similar in a limited way. A minor component of sand – titanium dioxide with some iron in it – can fix nitrogen. When the sun shines on damp crystals of this mineral small amounts of nitrogen are converted into ammonia, a form of nitrogen that is more easily built into bigger molecules such as amino acids.

Even under a rather inert atmosphere of mainly carbon dioxide, nitrogen and water vapour – a kind now generally favoured for the early Earth (pp. 5–6) – one can imagine the synthesis of small organic molecules taking place. Not on an ocean-wide scale by any means, but one can imagine local productions, for example where damp minerals were exposed to the atmosphere and to the early sunlight, rich in ultraviolet light. The amounts of materials produced would have been small, because such syntheses are, as far as we know, inefficient with ordinary minerals, and, as discussed in chapter 6, ultraviolet light destroys organic molecules too. Making and breaking (and jumbling up) would have been going on together.

And then there would be no escape from the more general problems associated with a purely geochemical synthesis of somewhat bigger organic molecules such as nucleotides. But this is no longer a critical barrier given mineral organisms. We are no longer talking about chance being able to bring about a long sequencing of procedures. We are asking now about what might be feasible given mineral crystal assemblages whose forms, associations, and defect structures could be contrived by natural selection. As discussed towards the end of chapter 6, when natural selection is in operation – when there is, in effect, a continuing memory of past successes – then such games of chance are radically transformed. Expertise can be gradually built up. The impossible may become highly feasible.

Could evolved mineral organisms be imagined as creating conditions for the synthesis of difficult molecules? We will come later on to the question of why they might do this. In the meantime let us stay with 'How?' questions. Is clay a suitable sort of material for the job? Is it the kind of stuff you would want?

The answer is 'yes'. Consider photosynthesis. This depends above all on a device for catching the light and on a device for keeping very small objects apart on a very small scale. (The electron and positive charge generated by the light package, and the first molecules made by these active agents, must be prevented from coming together again and cancelling each other out.) Iron atoms are common constituents of clay minerals, and ideal light-catchers, while the immensely thin and tough clay layer itself seems ideal as a micro-separator. The rest would be largely organisation. It would be a question of having the light-catching atoms appropriately placed, of having the clay membranes arranged in the right way, of having the shapes and sizes of the particles right – and so on.

Then the problems of sequencing, inherent in long organic syntheses, may also be solved in principle through appropriate spatial organisation. If this is not immediately obvious it is because spatial organisation is not the main technique that we as humans use to sequence manufacturing procedures. We rely rather on recognition. If we are baking a cake we do not have to have all the ingredients physically lined up in the order in which we are going to use them. Nor is it necessary for the spoons, mixer, oven, etc., to be precisely pre-aligned. Apart from anything else it would not do if our kitchen was set up only for making cakes. By contrast a one-product automatic factory will have much more of the manufacturing procedure built into its lay-out, relying little if at all on recognition.

In many ways the modern bacterium is more like a kitchen than a one-product automatic factory – it relies very much on the ability of enzymes to recognise other molecules. This is not as sophisticated as our human powers of recognition, but it is nevertheless a highly sophisticated ability. Early organisms could not have been like this. They would have been more like factories than kitchens. Those that came to be able to make molecules such as nucleotides would have been much more complicated than modern bacteria in the apparatus that they contained, the lay-out of which defined their set manufacturing procedures.

To be complicated is not, of course, the same as to be sophisticated. I ask you to imagine something like a complicated piece of laboratory glass-ware – a complicated collection of simple flasks, tubes, pumps, etc., connected together so as to define a particular sequence of operations. (Heath Robinson had the sort of idea – or, for the North Americans, Rube Goldberg.)

If you start thinking in this way, about the kind of 'glass-ware' that would have been needed for organisms embarking on the difficult field of organic synthesis, then clay minerals seems ideal materials. They can act as catalysts, but they are not too reactive; even without any kind of genetic control there are many clays that will form tubes or vessels. And many will hold on to organic molecules – on the edges of layers or, very often, stacked between them.

How could such apparatus have been put together under genetic control? One can imagine construction systems analogous to the folding of protein molecules. Perhaps those flexible clay layers with specific charge patterns written in them would crumple up or come together in specific ways so as to define a particular complicated piece of apparatus. Or perhaps grooved surfaces would come together leaving a complicated but contrived maze of fissures between them. Perhaps when we have made crystal genes of various sorts in the laboratory we will be able to be more specific about such 'How?' possibilities.

So let us move on to 'Why?' questions. What use could organic molecules be to evolving mineral organisms?

There are uses to be imagined at all levels, from simple poorly controlled mixtures that might act as glues or ultraviolet shields or crude barriers, to elaborate self-locking polymer molecules for more sophisticated purposes. And there would be many ways in which organic molecules could help in the processes of *making* clays.

Consider one of the simplest and most easily made organic molecules – formic acid. This might operate to stabilise the acidity of solutions and so help to control clay crystallisations. The slightly more complex molecule oxalic acid is also fairly easily made and is known to help clay synthesis, as are several of the central sub-component molecules that we discussed in chapter 7 (pp. 53–4). These kinds of molecule help clay synthesis by holding aluminium ions in solution.

Amino acids, and small strings of amino acids, are among other kinds of molecules that are good at holding on to metal ions in solution. Perhaps this is why amino acids were brought into a biochemical system in which amounts and concentrations of metal ions in solution were critical.

Moving out to the next shell of the supply structure (p. 55), the letter pieces of nucleotides are molecules of a sort that easily stick between clay layers, while the triple chain of phosphate units present in primed nucleotides (shown in appendix 1) is particularly good at sticking to the edges of clays. Perhaps molecules like nucleotides were designed in the first place to interact with clays, finding a first use in tying clay crystals together in special ways.

Then when we come to much bigger molecules we can imagine other uses. Polysaccharides are made from many sugar units joined together. They find various uses in modern organisms. Suitably made they become experts at controlling the consistency of solutions, making slimes of just the right sliminess, or jellies that will soften or harden under appropriate conditions. That is the kind of expertise that evolving clay organisms might find a use for – especially where success may depend on being able to solve such problems as staying in the light while not being dried out or washed away...

Perhaps the precursors of DNA and RNA were fancy poly-saccharides. In any case you can take it that they would have had no genetic use to begin with. The discovery that this class of molecules could replicate information would have come later – and it would have been an accident.

But how much more plausible now would be the accidental discovery of a replicating organic molecule, now that there were fully working organisms that had uses for many kinds of organic molecules and a gradually evolved expertise to carry through long organic syntheses?

My guess is that the reasons why RNA-like chains first took hold were purely structural reasons, depending on the ability of different segments of chain to lock together (see figure on p. 23). This was a way of making complicated objects – a construction system. But given the establishment of such a technique the possibility of replication would suddenly be there.

We discussed this *sort* of thing in chapter 8. It happens all the time in evolution: something evolved for one purpose often – *usually* – turns out to have other uses (remember the cat's tongue). There is no question in such cases of fore-thought; only a kind of opportunism.

Once there was another set of replicating structures able to hold and pass on information then evolution might take place through (replicable) changes in this new material as well as the old. The required condition would be that sequence information in the RNA-like molecules should have some effect on a property that was of some use to the clay-organic organisms as a whole. For example, the replicating sequence might determine the way in which the molecules folded up on themselves to make useful pieces of machinery, or locked with each other to provide a well-tuned structural material.

At least some amino acids would have been present within the evolving organisms by now, and we must suppose that, somehow, well-organised RNA-like molecules came to help amino acids to join up into chains. This must have been a long gradual process (and *surely* only possible within fully working, evolving organisms). Only when our central control machinery was complete, genetic code and all, would there have been any prospect of free-living organisms based exclusively on this marvellous new system.

Indeed there was still more to be done: protein enzymes had to evolve as well as altogether new kinds of membranes etc. to replace that clumsy clay apparatus. Eventually it was all replaced. The system was then stuck more or less as we see it now. Nothing could compete with this slick super-life made largely from air and sunshine. (Look, no clay!) One particular species of it became the common ancestor of all life now on the Earth.

Why did organic molecules win such a resounding victory? That question was discussed in chapter 9. Organic molecules are better for 'high-tech' machinery. This is mainly because the atoms in them are held together more securely. These atoms do not self-assemble at all well: it is much more difficult to make coherent orderly multi-atom structures containing carbon than to make, say, such silicon–oxygen structures: but, once made, carbon-based structures can retain their individual complexity indefinitely. By contrast a crystalline structure formed from water solutions is always in danger of dissolving away again in water; and very small crystals, or small pieces of a crystalline object, are particularly liable to fall apart or re-arrange. It is the other side of the same coin: if you can self-assemble easily, you can self-disassemble easily too. This is alright for 'low-tech' perhaps, but it is limited. You need more than sticks and string for serious engineering.

So in the end the supremacy of organic bio-materials is tied in with the question of scale. Organic machinery can be made much smaller. Such clever things become possible as sockets which can recognise, hold and manipulate other molecules – and in any competition to do with molecular control the system with the smallest fingers will win.

'It appeared to you to be a simple case; to me it seems exceedingly complex.'

15

Summing-up: The seven clues

'...the special points upon which the whole mystery turns.'

First clue: from biology

Genetic information is the only thing that can evolve through natural selection because it is the only thing that passes between generations over the long term. Although held in a genetic material, genetic information is not itself substance. It is form. But it is a sort of form that can outlast substance, because it is replicable. Evolution can only begin once there is this kind of form – when the conditions exist for the replication of genetic information.

This first clue was by far the most important. It directed our attention to the real issue: it set the question. And it was to suggest an answer, if only in general terms, as to what the very first organisms must have been like. They must have been 'naked genes', or something close to that. This clue first appeared early on – *in chapter 2*.

Second clue: from biochemistry

DNA is a suburban molecule far from the centre of the present biochemical pathways. The same can be said of RNA. Biochemically as well as chemically these are evidently difficult molecules to make: it takes many steps to manufacture even just their nucleotide units from the simpler central molecules of biochemistry. All this suggests a comparatively late arrival for these now undisputed rulers.

The second clue seemed to be in conflict with the first, which had clearly implied a working genetic material from the start. (The resolution of this conflict was to be the way forward.) The second clue was a long time in coming; there had been some detailed facts to be presented, and two red herrings to be removed first. Foreshadowed in chapter 6, this clue finally appeared *towards the end of chapter 7*.

114

Third clue: from the building trade

To make an arch of stones needs scaffolding of some sort; something to support the stones before they are all in place and can support each other. It is often the case that a construction procedure includes things that are absent in the final outcome. Similarly in evolution, things can be *subtracted*. This can lead to the kind of mutual dependence of components that is such a striking feature of the central biochemical control machinery.

This third clue alerted us to the likelihood of a missing agent, an earlier 'scaffolding' – an earlier design of organism at the start of evolution. And it seemed very possible that these first organisms would have been based on a genetic material no longer present *at all* in our biochemistry. The clues were coming thick and fast now: this one *early in chapter 8.*

Fourth clue: from the nature of ropes

None of the fibres in a rope has to stretch from one end to the other, so long as they are sufficiently intertwined to hold together sideways. The long lines of succession that alone connect us to distant ancestors are like multi-fibred ropes in that what are passed on between generations are *collections* of genes ('intertwined' because they correspond to viable organisms and it is thus in their mutual interest to stay together). But new 'gene fibres' may be added and others subtracted without breaking the overall continuity.

This fourth clue was about ways and means. It suggested to us how organisms based on one genetic material could gradually evolve into organisms based on an entirely different genetic material. This was the central clue, in more ways than one. It appeared *in the middle of chapter 8.*

Fifth clue: from the history of technology

Primitive machinery is usually different in its design approach (and hence in materials of construction) from later advanced counterparts. The primitive machine has to be easy to make from immediately available materials; and it must work, more or less, with minimum fuss. The advanced machine simply has to work well, but it does not have to be particularly easily assembled: it can be made from diverse specialist components working in collaboration – and usually is.

This fifth clue led us to suspect that the first, unevolved (necessarily 'low-tech') organisms would have been very different from the (manifestly 'high-tech') organisms of today. Most probably their materials of construction would have been very different too. This clue appeared *late in chapter 8*.

Sixth clue: from chemistry

Crystals put themselves together, and in a way that might be suitable for 'low-tech' genetic materials. Even the most primitive kind of gene-printing process would have to be fairly precise and involve the coming together of a fair number of atoms. Big organic molecules show little sign of having the appropriate self-control. On the other hand there are several cases where the replication of complex information can be imagined as taking place through crystal growth processes.

The sixth clue gave a sense of direction to our search for primitive biochemical materials. The significance of this clue emerged gradually through chapter 9 – and it took much of the rest of the book to develop. It first came into sharp focus *at the start of chapter 10*.

Seventh clue: from geology

The Earth makes clay all the time, as you can see from the huge amounts of it that are carried in rivers. The minerals of clay are tiny crystals that grow from water solutions derived from the weathering of hard rocks. Not only for primitive genes, but also for other primitive control structures such as 'low-tech' catalysts and membranes, these kinds of inorganic crystals seem to be much more appropriate than big organic molecules.

The seventh clue depends for its significance on all the others. It is certainly no new idea that this most earthly of materials, clay, should have been the stuff of first life – it is in the Bible. What *is* new is our understanding of just how interesting, varied and complicated this sort of stuff is when looked at under a super-powered magnifying glass. The seventh clue appeared *in chapter 11*.

These, then, are my seven best clues to the origin of life. Only the first could be said to represent an important insight in itself (and it is by no means new); and only the second is at all technical. The others are commonplace. But then as Holmes said:

'The interplay of ideas and the oblique uses of knowledge are often of extraordinary interest.'

APPENDIX 1

Units for DNA and RNA

Here is the way in which the atoms are joined to each other within one of the four DNA nucleotides:

(A)

This molecule can be thought of as having two halves: there is a 'letter piece' (shown in heavy type on the right), and a 'connector piece' (on the left). The DNA nucleotides all have the same connector piece, but the letter pieces may be either adenine (A) as shown above, or guanine (G), thymine (T) or cytosine (C):

(G)   (T)   (C)

To make a DNA chain the phosphorus atoms (top left) in the connector pieces of the nucleotide units have to join up with oxygen atoms (bottom left) of other nucleotide units. In effect water molecules have to be pushed out – in two pieces, OH coming off the top and H coming off the bottom, as indicated in the picture. (This is to release appropriate 'press-studs': see p. 17.)

In practice nucleotide units are not able to link together in this way unless they are 'wound-up' or primed first. This is done by replacing the (too comfortable) OH piece with something that will break off much more easily – a short string of additional phosphorus and oxygen atoms. Here then are the structures of the four primed DNA nucleotides:

The four RNA nucleotides differ from the DNA nucleotides only in having a slightly different connector piece (with one more oxygen atom in it) and in having a modified version of the letter piece thymine (without the CH$_3$ knob) called uracil (U).

Units for proteins

Here are three of the twenty kinds of protein amino acids:

alanine

glutamic acid

tyrosine

Like nucleotides, the amino acids can be thought of as being made up of (this time rather small) connector pieces as well as (this time very varied) letter pieces. And the mode of linking up is similar, in that the formal process requires the breaking off of OH and H (as indicated above). Again in practice these links have to be primed first, although the priming process is more complicated in this case.

APPENDIX 2

The kaolinite layer

The picture opposite represents a very small part (about a millionth) of only one layer (out of thousands) making up a kaolinite crystal (that would still be a hundred times too small to be visible to the naked eye).

So as to be able to see into the structure the atoms are shown proportionately much smaller than the distances between them. The lines represent covalent bonds. The layer is represented as lying flat on the page and consists of five planes of atoms (not including hydrogens) at five different levels. The top plane is of hydroxyl groups (oxygen atoms with a hydrogen attached to each). These hydroxyl groups are represented by bull's-eyes in heavy type. Next down is a plane of aluminium atoms (very small white circles) below which is a mixed plane of simple oxygen atoms (larger white circles) and some more hydroxyls. Then below that there are silicon atoms (very small black circles) and below that again a plane of only oxygen atoms (the more faintly drawn white circles).

If you look carefully you will see that while a silicon atom is always surrounded by four oxygen atoms, an aluminium atom is always surrounded by six oxygens. The silicons are said to occupy fourfold, and the aluminiums sixfold sites. If you look more carefully still you will see that only two-thirds of possible aluminium sites are in fact occupied (one of the vacant sixfold sites is shown by dashed lines) and that the vacant sites all lie at the same angle in relation to a hexagonal background pattern created by the silicon atoms and the oxygens that lie beneath them. If you are not quite cross-eyed by now you will see that this direction is at 'one o'clock'. It could equally have been drawn at 'five o'clock' or at 'nine o'clock'. What matters most is that there *is* a direction to this complex patterning of atoms,

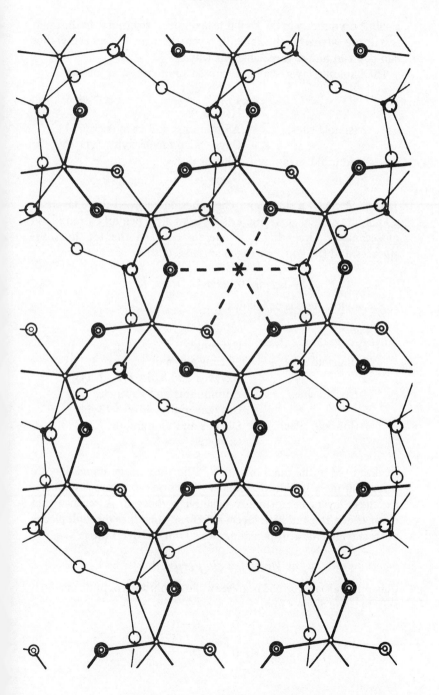

a subtle asymmetry in the kaolinite layer that I referred to in the main text as the 'arrow' in the kaolinite layer structure. It is an arrow that can point in one of three different ways.

The kaolinite layer structure can be summarised as follows:

Plane	
1	Hydroxyls
2 (sixfold sites)	Aluminiums and vacancies (2:1)
3	Oxygens and hydroxyls (2:1)
4 (fourfold sites)	Silicons
5	Oxygens

This is, in effect, a side view. The cohesion between these layers in a kaolinite crystal is strong, arising from a particularly strong kind of secondary force that can operate between the top hydroxyl surfaces and the bottom oxygen surfaces of these sorts of layers.

The (ideal) muscovite mica layer

The side view here is as follows:

Plane	
1	Oxygens
2 (fourfold sites)	Silicons and aluminiums (3:1)
3	Oxygens and hydroxyls (2:1)
4 (sixfold sites)	Aluminiums and vacancies (2:1)
5	Oxygens and hydroxyls (2:1)
6 (fourfold sites)	Silicons and aluminiums (3:1)
7	Oxygens

As discussed in the main text (p. 83) the aluminium atoms that are present in place of silicon atoms in the fourfold sites are responsible for these layers being highly negatively charged. A mica crystal consists of a stack of such layers held firmly together through planes of (positively charged) potassium ions lying between them.

Units for clay crystals

Clay minerals are formed from very dilute solutions of units of which silicic acid is always one:

$$\begin{array}{ccc} H-O & & O-H \\ & \diagdown Si \diagup & \\ H-O & \diagup \quad \diagdown & O-H \end{array}$$

Leaving out the hydrogen atoms this can be represented as:

These units can link up, pushing out water molecules in the process:

Such linkings take place easily, as do unlinkings – that is to say the processes are reversible. The bonds being formed and broken are nevertheless strong covalent bonds. The secret behind these seemingly contrary statements is *exchange*. If one strong bond can be exchanged for another through appropriate comings and goings of water molecules then these bonds will be in effect quite labile. But once out of contact with water, deep inside a crystal, bonds like these are no longer easily made or broken.

The other units needed to make clay crystals are hydrated metal ions. These are positively charged metal atoms with water molecules (usually six) around them. Ignoring the charges they can be represented as:

or:

These too can join together in a reversible way to make bigger strongly bonded structures:

Or silicic acid units may join up with hydrated metal ions:

All such comings and goings allow particularly stable arrangements to be discovered – and then these tend to persist. (This is typical of any sort of crystallisation process.) Kaolinite and mica-type layers represent such particularly stable arrangements. (Look again at the picture on p. 121.) It takes billions of joinings (and unjoinings) for even the tiniest clay crystal to put itself together.

GLOSSARY

The purpose of this glossary is to remind the reader of the meanings of terms already given more or less explicitly in the text (page numbers in brackets). The glossary is in two parts. The first part relates to *the nature of the problem* of the origin of life. It recapitulates key terms that are introduced in the first four chapters. This part can be read early on. The second part gives away the plot to some extent – so if you want to work out for yourself the way the arguments are leading, it might be as well to leave this part till later. In any case this second part recapitulates terms that appear after chapter 8 relating to *the nature of the solution* to the problem of the origin of life that is being put forward.

Part I

An **organism** (3) is that which can take part in the processes of **evolution** (1–4) through **natural selection** (2–3). For this it must have a dual constitution, namely: (i) a store of **genetic information** (9) or, as we have been calling it, a **Library** (9–10). This aspect of an organism, its genetic constitution, may comprise more or less individual items of information, conveniently called **genes** (12) – a term that we have been using somewhat informally. The other aspect of an organism is (ii) its **phenotype** (9) – its outward and visible parts, the effect or expression of its genetic information.

Life (1–3) is an informal term for the seemingly purposeful quality of evolved organisms. If organisms are prerequisites for evolution, 'life' is rather a *product* of that process.

Atoms (16) can be regarded as the building bricks of materials on the Earth. A **molecule** (16) is a group of atoms joined together in a particular way by **covalent bonds** (17). An **ion** (17) is an atom or molecule with electric charge. **Organic molecules** (17–18) are ones that have carbon atoms in them. Forces that are weaker than

125

covalent bonds – **secondary forces** (20) – operate between molecules which, like all very small objects, are in continual haphazard motion – what we have been calling **heat agitation** (21).

Proteins (24–6) are large organic molecules and key constituents of phenotypes of modern organisms. Among these there are many, such as the **enzymes** (20), that can manipulate other organic molecules. Proteins are built from a set of twenty fairly small molecules called **amino acids** (25, 118–19) under the control of genetic information in the **genetic material** (22), which is (today) DNA (22–4, 117–18). RNA (27, 118) is another very similar **nucleic acid** (27) which helps in making protein. DNA and RNA are built from molecular units that are larger and more complex than amino acids and called **nucleotides** (22, 117–18) although, like the amino acids, these have to be **primed** (23, 118) or 'wound-up' before they will link together.

Part II

A **naked gene** (66–7) is a hypothetical minimal organism that has no separate phenotype.

Atoms and molecules often undergo **self-assembly** (68–71) into higher-order structures under the combined effects of their heat agitation and of **reversible** bonds (70–1), or forces, acting between them. These are bonds that can be readily made and unmade in given circumstances. **Crystals** (70, chapter 10) are the commonest class of self-assembled objects. **Crystal genes** (74, chapter 12) are hypothetical primitive genes replicating through crystal growth processes. Ordinary crystals can grow in solutions which are sufficiently rich in suitable units – which are **supersaturated** (75–6) – although such solutions often have to be **seeded** (75–6) first with a piece of the crystal that is to form.

A crystal is usually made from vast numbers of units (atoms or molecules, often ions) packed together for the most part in some regular way described as its **crystal structure** (76). But there are always irregularities of packing in real crystals – so-called imperfections or defects – which give an individual crystal a more particular **defect structure** (88). **Twinning** (90–2) is a common form of defect where different regions in a crystal are misaligned. **Substitutions** (83) are a form of defect in which 'wrong' units are incorporated in a crystal structure in a more or less haphazard way.

Clay minerals (chapter 11, 120–4) are composed of exceedingly minute crystals, typically **layer silicates** (81), of which **kaolinite** (81,

120–2), **dickite** (82) and **halloysite** (82) have one sort of layer in them; while **illites** (83) and **smectites** (83) have another – similar to the layers in the mica mineral **muscovite** (82–3, 122). **Polytypes** (82) are crystals that differ in the way in which identical layers are stacked on top of each other. Such stacking modes may be orderly, but there are also **disordered polytypes** (84). Different kinds of layers may also be stacked together within what is then called a **mixed layer** crystal (84).

We have been assuming that the first organisms on the Earth arose, not from pre-existing organisms, but through **spontaneous generation** (98). Being able to arise like this they would have been **primary organisms** (99) – unevolved and so not-yet-alive to begin with. Life would have been a later gradual emergence, and our kind of life, based on **secondary organisms** (107), would have been later still. These 'high-tech' secondary organisms would have evolved from primary organisms through a gradual replacement of genes made of one material by genes made of an altogether different material – that is to say through a **genetic takeover** (62–3, chapter 14).

INDEX

acetyl group, 54
adenine, 36, 44, 117
age of the Earth, 5
alanine, 46, 118
aluminium atoms in clays, 81–2, 83,
    120–3
amino acids, 19, 35, 36, 43–4, 46, 54,
    71, 111, 112, 126
  molecular structure, 118–19
  protein units, 25
ammonia, 4, 6, 36, 108
ancestry, common, of organisms, 41–2,
    52
anthropic principle of origin of life, 7
arabinose, 45
aspartic acid, 60
atoms, 16–19, 125
  motion, 21
ATP, 45

bacteria, 11, 14, 98
biochemical materials, the first, 63
biochemical system, nested, 53–7
biochemistry, unity of, 35, 37, 38–42
biology, 1

C-hypothesis, 55–6
caramel, 44
carbon atoms, 16, 17–18, 54
carbon dioxide, 6, 54, 56–7, 64, 108
Carter, Brandon, 7
cells, 9–10
  division, 13–14
  single, 11
chance
  organic synthesis and, 47–8
  origin of life and, 4, 6–7, 99
characteristics, inherited, 12
chemical defects in crystals, 92–3
chemical elements, 16

chemical evolution, 35–7
chlorophyll, 108
chromosomes, 10
clay crystals, 100–3, 105
  effects of organic molecules, 110–11
clay materials, 80–6, 126
  crystallisation, 80–2, 84–6, 122–4
  structures, 82–4, 120–2
clays, 116
  mixed layer, 84, 93, 94
code, genetic, 40, 61, 112
Coleridge, S. T., 2, 58
complexity of biochemical systems, 40
control structure of organisms, 53–7
conventions in biochemical systems,
    40–1
copying of messages, 13
covalent bonds, 17, 19, 71, 125
  between atoms, 71
Crick, Francis, 8
crystal genes, 74–5, 77–9, 87–97, 100,
    102, 105–6, 126
  replication, 84–96
crystal growth, 75, 90–1, 116
  continuous, 78–9
  illite, 102–3
crystal structure, 76–7, 88, 126
crystals, 70, 99, 126
  clay, 100–3, 105
  columnar, 89–90
  defect structures, 88–90, 91–3, 126
  growth, 75
  replication, 101–2
cues, for perception, 31
cyanide, 36, 43
cytosine, 117

Darwin, Charles, 2, 51, 58
defect structures of crystals, 88–93,
    126

128